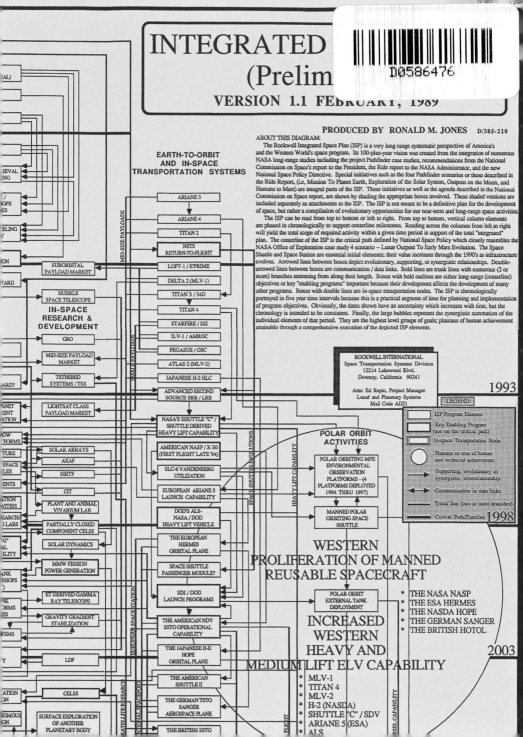

AD ASTRA

AD ASTRA:
AN
ILLUSTRATED
GUIDE TO
LEAVING
THE PLANET

DALLAS CAMPBELL

SIMON &
SCHUSTER

London · New York · Sydney · Toronto · New Delhi

A CBS COMPANY

First published in Great Britain by Simon & Schuster UK Ltd, 2017
A CBS company

1 3 5 7 9 10 8 6 4 2

Simon & Schuster UK Ltd
1st Floor, 222 Gray's Inn Road, London WC1X 8HB

www.simonandschuster.co.uk
www.simonandschuster.com.au
www.simonandschuster.co.in

Simon & Schuster Australia, Sydney

Simon & Schuster India, New Delhi

The author and publishers have made all reasonable efforts to contact
copyright-holders for permission, and apologise for any omissions or errors in
the form of credits given. Corrections may be made to future printings.

A CIP catalogue record for this book is available from the British Library

Hardback ISBN: 978-1-4711-6405-7
eBook ISBN: 978-1-4711-6406-4
eAudio ISBN: 978-1-4711-6749-2

Art Direction and Design: Ashley Western
Project Management: Nicola Crossley and Tamsin Edwards
Editorial: Mike Jones, Martin Bryant and Caroline Blake
Picture Research: Ian Whent
Text Permissions: Louise Crane
Illustrations: Bee Willey

Printed in Germany by Mohn Media Mohn Media GmbH

Previous page: **Neil Armstrong practises his small step**

CONTENTS

For James and Anna

and everyone who enjoys
staring into space

Earth in the viewport
Earth in the viewport
Earth in the viewport that I see...

Like a son missing his mother,
Like a son missing his mother,
We miss our Earth – we have but one...

But the stars nevertheless,
But the stars nevertheless,
Draw closer, though they're still as cold as ever...

And like at the hour of eclipse,
And like at the hour of eclipse,
We wait for light and see our earthly dreams...

And we dream not of thunder at the cosmodrome,
Not of the ice-cold blue of the sky –
But we dream of grass – the grass beside our house
Green, green grass...

And we fly our orbits,
Unbeaten paths –
Lifetimes like meteors in the vastness...

Courage and risk were justified,
For the music of space
Floats into our matter-of-fact talk...

In some opaque haze
Earth in the viewport –
An early evening-time twilight...

But the son misses his mother
But the son misses his mother –
The mother waits for her son, as the earth awaits her children...

And we dream not of thunder at the Cosmodrome,
Not of the ice-cold blue of the sky –
But we dream of grass – the grass beside our house
Green, green grass...

Zemlyane ('Earthlings')
'The Grass Beside Our House' (1979)

PREFACE

15 DECEMBER 2015

THE BAIKONUR COSMODROME
KAZAKHSTAN

45.965° N, 63.305° E

If you've never heard of the Soviet 1980s synth-rock band Zemlyane ('Earthlings') and you're thinking about planning a trip to space, it might be worth checking them out. If you happen to be going into space *today*, it's mandatory. The song 'The Grass Beside Our House' is a rousing, hair-sprayed soft-rock anthem that's played to the astronauts as they emerge from the hotel to begin the launch day's preparations. You are not allowed to fly on the Soyuz rocket unless that song has been played.

I'm standing outside the hotel in a big press scrum in the town of Baikonur, a Russian enclave located by the vast Cosmodrome rocket range in the desert steppe of Kazakhstan. It's from this place, in what was the Soviet Union, that the story of leaving the planet began, on 12 April 1961, when Yuri Gagarin launched from here to orbit once around the earth aboard Vostok 1. Here we are, nearly sixty years later, in the same place, about to watch a launch from the same launch pad on what is essentially the same rocket.

It's early morning and well below freezing. The throng of well-wishers, families, security and press are ready to greet the crew as they emerge. 'The Grass Beside Our House' is playing full blast, distorted through the ancient loudspeakers. I spot a few plastic Union Jacks being waved, and some pre-emptive shouts of 'Good luck Tim!' The press officer from the European Space Agency (ESA)

looks bemused. He hasn't seen anything quite like this before: 'Why is everyone *so* excited?' It's a very good question.

As Zemlyane's guitar solo builds to its climax, the doors of the hotel open, the cameras spring into action and the three astronauts appear, waving to the cheering crowd, flanked by officials, doctors and the imposing black-robed, bearded Russian priest who yesterday had blessed the Soyuz TMA-19M rocket with holy water in its launch position, pointing towards the heavens. 'The holy water gives it a little more va-va-voom...' we journalists joke, congregated together on the launch pad, also receiving a blessing with a generous soaking. Soyuz crews have a long list of superstitious rituals that need to be ticked off: last night they and their families sat down to watch the film *White Sun of the Desert*. Every astronaut that has flown on Soyuz since 1971 has watched this film the evening before launch. Then there's the 'planting of the tree', the 'signing of the door' and today, most famously, 'the peeing on the rear right tyre of the bus that takes the astronauts to the launch pad'. Female astronauts, if they wish, can carry a little specimen jar of wee (or they can opt out). It's a bit of a faff otherwise, what with the spacesuit zips and everything. Any parent who's begun a long car journey with kids knows the form. Yuri Gagarin stopped the bus because he (sensibly) just had to go, and if that's what happened on his launch day, that's what happens now.

The three astronauts – Commander Yuri Malenchenko, a steely eyed missile man and veteran of six space flights; Tim Kopra, the American NASA flight engineer, who doesn't seem to mind being 'the other Tim', for today at least; the man of the moment, Major Tim Peake, second flight engineer, the first British ESA astronaut, who is grabbing all the headlines in the UK – have a busy morning ahead of them. Together they are Expedition 46/47 and are about to spend six months on board the International Space Station (ISS), our most remote human outpost and the furthest we've been* since the final Apollo 17 expedition to the moon in 1972. Like 1980s rock stars in their mission-patched blue jumpsuits, they make their way through the crowd to board the coach which, disappointingly, is the only non-Soviet era thing around.

In Baikonur it will forever be 1961. Wherever you go around the town there are reminders of its extraordinary place in space history. The local cafe has pictures of every crew who have flown from here; wild Soviet space murals decorate the sides of buildings; half-size mock-ups of Russian rockets line the streets, with statues of the men who built them. Most significant is the bust of Sergei Korolev, the architect of the R-7, the world's first intercontinental ballistic missile (ICBM), designed to throw a nuclear warhead across the globe onto American soil but, as luck would have it, also really good at throwing satellites, dogs and humans into orbit. Its latest incarnation, the Soyuz launch vehicle, has been the workhorse of human space flight for decades. Since the retirement of the American Space Shuttle in 2011, the Baikonur Cosmodrome is the only bus stop to space for all space-faring nationalities, other than the Chinese who are marching ahead with their own human space programme. And Soyuz, beautiful reliable Soyuz, 'queen of the skies', is the only vehicle that can get you there.

We are in a transitional period. The broken concrete, rust and peeling paint of the Cosmodrome represent the past and the present. But great changes are afoot. Shiny new spacecraft and launch vehicles are being developed: NASA's unimaginatively named Space Launch System (SLS) promises to get us on the road to Mars. Elon Musk's SpaceX is launching spectacular self-landing reusable rockets, and promising commercial trips around the moon – all part of his grand plan to propel humans to becoming a multi-planet species. Companies like Blue Origin and Virgin Galactic are opening the doors to sub-orbital roller coaster rides. The moon and Mars are back in the sights of ESA and NASA. The rate of exciting new space announcements is increasing exponentially. For those who have cosmic wanderlust, these promise to be exciting times.

At the launch pad, called Gagarin's Start, there's no giant digital display or Thunderbirds-style countdown. I'm standing with the various TV crews right up against a rusty barbed-wire fence a little under a kilometre from the launch pad, doing a live link for the BBC. A small stray dog, one of many that live around here, is curled up asleep next to a rusting football goalpost behind me. The dogs that haunt this place have seen all this before of course. There is no advanced warning that the rocket is about to leave – you just have to concentrate. The parallel arm of the service gantry pivots eerily away from the rocket as if by telekinesis and a few moments later you hear the hiss of the fuel injectors, followed by a low rumble, exactly like a rumble of thunder, that grows to a deafening roar. It's one of the most beautiful sounds you'll ever hear. Like a flower opening, the simplicity of the launch system reveals itself – as the rocket becomes weightless, Soyuz is released like a firefly, her flame too bright to look at directly. Three out of 7.3 billion humans are leaving our planet,

---

* Other than the Hubble Space Telescope Servicing Missions, which were in a higher orbit.

into the ice-cold blue sky and through to the other side. In a few hours the population of our off-world colony, the International Space Station, will have doubled to six.

Have you ever imagined joining them? If so, your perception of what it might be like will be formed from a rich broth of stories, images and ideas pulled from science fiction and historical fact, two threads that have always been woven together. You'll almost definitely have seen *2001: A Space Odyssey*, *Star Wars*, *Apollo 13* and a thousand other films, TV shows, cartoons and documentaries too. You might have read or seen Carl Sagan explore the universe in *Cosmos*, or read Tom Wolfe's *The Right Stuff*, chronicling the early years of the American space race – all silver spacesuits, Corvettes and Chuck Berry on the jukebox. Perhaps you're of an age to remember where you were when you first saw the grainy live TV pictures of Neil Armstrong stepping onto the surface of the moon. Planting flags on the moon was what Americans did back then. Today, nearly fifty years later, we wonder at the scale and ambition of such a project, and in our risk averse world, can scarcely believe it ever happened.

Going into space isn't easy. At the time of writing, 553 humans have done it. Only 24 people have gone beyond earth's orbit, and of those, 12 of them have walked on the moon – of whom 6 are still with us. Seven people have paid to go into space with their own money. Eight people have died en-route to space and 11 on the way home, with several others losing their lives in training.

The doors to this exclusive club are opening to new members soon. If you want to be astronaut number 554 – or if you simply want to join the 108 billion* who have yearned to look back at the earth, or imagined setting foot on other worlds, this book will hopefully whet your appetite and help you on your way.

---

* Everyone who has ever lived.

FIRST STAGE:
GROUND CONTROL

Поехали!
Yuri Gagarin

06:07 UT, 12 April 1961

LET'S GO* on a journey into space. If you set sail from the earth and kept going, what would you see? Where and when would you end up? Let's leave the cradle and head to the moon. First we pass through the satellites and debris that circle the earth. A thousand pairs of eyes look down, and junk born from junk ad infinitum. Let's go into circumlunar space. We watch the moon getting bigger, and the earth receding behind us to a small blue marble. Above the atmosphere the stars around us are still – space travel takes the twinkle out of stars. Fixed points in a black sky. As we speed up, we pass some of our familiar planets – Venus, Mars, Saturn. Much further than we've ever been.

Let's go up away from the sun. Through the Kuiper Belt of dust and ice left over from the formation of the solar system that lies beyond the gas giant Neptune to where Pluto resides. Further still, and out past the lapping shoreline of the heliosphere, where the solar wind dies away and into the deep waters of the interstellar medium – the void between the stars. Here, punctuated by dense clouds of molecular hydrogen, are the foundations of the stars yet to be born, blown by stellar winds. We would not see them with our eyes, which are tuned like a radio to visible light, but they would be there. Turbulent molecular clouds of gas blotting out the star light.

Further still, *ad astra*, and we reach our first star – Alpha Centauri. A realm of stars and planetary systems like our own. Further still above the plane of our own Milky Way galaxy, rotating slowly like a giant illuminated sombrero, is the huge bulge of stars in the centre. Behind us another almost identical galaxy, Andromeda, and around us Magellanic clouds, and the dwarf galaxies of Ursa Minor and Fornax. Like the planets in our solar system, these are our Local Group of galactic neighbours. In 4.5 billion years, Andromeda will collide with us. And yet for all their billions of stars, the two galaxies will pass through each other without a single collision, the stars dwarfed by the distance of the spaces separating them.

Further back in space and time we enter the intra-supercluster medium, part of our own Virgo Supercluster of galaxies that make up the unfashionable end of the protruding filament of the Laniakea Supercluster, from the Hawaiian word meaning 'immeasurable heavens'. At the centre of Laniakea, galaxies flow towards an area called the Great Attractor like a bath plug tugging at the water. We can see the structure of the universe. We are in a realm of filaments and voids – the Sloan Great Wall of galaxies, over a billion light years across.

Let's go back further still to see the cosmic knitted web of matter, dark and light, like a crocheted blanket now surrounding us, and beyond the opaque *terra incognita* that takes us to the edge of the map and back to the beginning...

Opposite: **The universe**

---

* Yuri Gagarin's words as he left the planet.

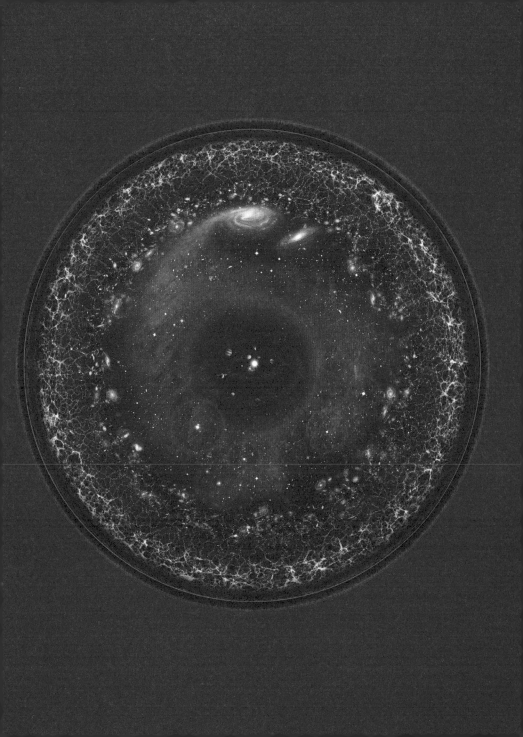

WHAT'S HOLDING YOU BACK

THE JACOBEAN SPACE PROGRAMME*

If you're old enough, you might just remember where you were when the history of human spaceflight began – that seminal moment in 1638 when Domingo Gonsales, a diminutive Spanish adventurer, left the island of Tenerife and flew to the 'Moone', in English Bishop Francis Godwin's 'picaresque' tale of space travel, *The Man in the Moone*. Gonsales made his lunar voyage not on a rocket, but by harnessing a flock of a special breed of migrating lunar geese called *Gansa*, that pulled him across the circumlunar space between the earth and the moon in some twelve days. Equipped with a religious certainty and a new scientific questioning, he lived for many months with the native *Lunars*, before heading back to earth. The tale is a reflection on the practicalities of space travel, extraterrestrial life, the universality of Christianity and the exotic new science ideas of the day: planetary science and orbital mechanics, and early ideas about gravity and electromagnetism. It's a sort of Jacobean *James and the Giant Peach* and essential background research for anyone wanting to leave the planet. Perhaps we can call Domingo Gonsales our first proto-astronaut? A 'celestial Don Quixote'? A missionary Buck Rogers?

Godwin wasn't alone. The French writer Cyrano de Bergerac also imagined *A Voyage to the Moon* in 1657, propelled to the heavens not by goose power, but by an altogether different scientific phenomenon – the evaporation of dew. Reasoning that the dew on the ground was attracted by the sun's rays, he bottled it up, strapped the bottles around his body, creating a sort of Jacobean jet pack, rising high above the clouds, but a few broken bottles meant his weight finally brought him back to earth.

In the same year that Francis Godwin's lunar geese story was published, John Wilkins, a founding member of the Royal Society, wrote his own altogether more practical thoughts on lunar travel in *The Discovery of a World in the Moone*. The book's cover illustration shows the sun being orbited by the earth and moon, an advert for Copernicanism. Like Godwin's adventure, behind the story was a popularization of the new science du jour. Above all, Wilkins was fascinated by the practicalities of the mechanical world, writing about catapults and levers, springs and gears in his later work *Mathematical Magick* (1648), as was David Russen in his all but forgotten *Iter Lunare* (1703) which entertained the idea of getting to the moon by using a giant, powerful spring.

These lunar imaginings didn't spring from the ether. The seventeenth century, much like the second half of the twentieth century, was an important period for 'Moone shots'. It was the time where our

---

* Title from Professor Allan Chapman. See the end notes section for information on sources and other thoughts.

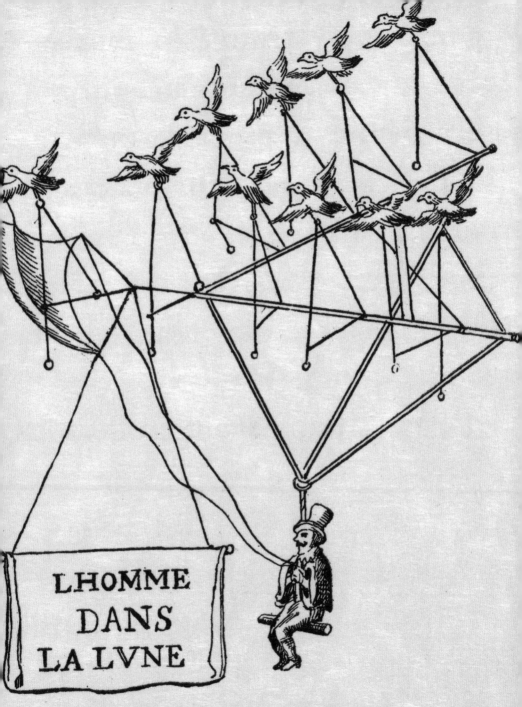

LHOMME
DANS
LA LVNE

Frontispice d'un roman de Gonzalès. — 1648.

understanding of the universe and our place in it were slowly being turned inside out. The classical world of Ptolemy and Aristotle was giving way to ideas grounded in new technology, exploration, empire-building, art and a European scientific revolution, underpinned by our most important invention of all – the scientific method.

THE RENAISSANCE RIGHT STUFF

In the fifteenth and sixteenth centuries, the world had been gradually mapped by the great voyages of Columbus, Magellan, Drake and Da Gama. Their sailing ships had pushed beyond the horizons, revealing the planet as it was rather than how it had been imagined. Europe was no longer the centre of the universe. Indigenous people, speaking in strange tongues, worshipping blasphemous gods, were discovered, often with disastrous results. Commerce and trade routes were being established, providing a new motivation to explore.

Similarly, the radical new invention called the telescope permitted us for the first time to reveal the celestial bodies, not as imagined supernatural manifestations, but as viable destinations. Solid islands floating in the ocean of space. Galileo Galilei, looking to the heavens with his new configuration of lenses, confirmed the Copernican model of the solar system as *heliocentric* rather than *geocentric*, not just in theory but in testable practice. For the first time the moon's surface was revealed. Planets and their moons were drawn. Sunspots were seen moving across the face of the sun. The architects of this revolution in thinking – Tycho Brahe, Robert Boyle, Francis Bacon, Johannes Kepler, Robert Hooke, John Wilkins, and later Isaac Newton – were studying God's intricate handiwork in detail. Gravity, magnetism, astronomy and anatomy were all under the spotlight.

In the seventeenth century, this new knowledge seeded the imagination of the scientists themselves. In 1608, physicist Johannes Kepler, famous for his work describing the orbits of the planets, imagined in his novel *Somnium* what the earth would look like from the moon. He describes the logistics of a moon journey propelled by supernatural 'daemons', with a detailed description of the physiological challenges of space travel to humans. Like Arthur Dent and Ford Prefect's ten pints of beer taken as 'muscle relaxants' in the pub before their ride on the Vogon ship in *The Hitchhiker's Guide to the Galaxy*, Kepler's astronauts are prescribed drugs and opium for their space voyage to combat the physical shock. Lack of air and cold were to be remedied by cold sponges up the nose.

What kind of people would Kepler sign up as space travellers? Here are his thoughts so you can see how you measure up:

We do not admit desk-bound humans into these ranks, nor the fat, nor the foppish. But we choose those who regularly spend their time hunting with swift horses, or those who voyage in ships to the Indies, and are accustomed to living on hard bread, garlic, dried fish and other abhorrent foods. The best adapted for the journey are dried-out old women, since from youth they are accustomed to riding goats at night, or pitchforks, or travelling the wide expanses of the earth in worn-out clothes. There are none in Germany who are suitable, but the dry bodies of Spaniards are not rejected.

TRUE STORIES

> My subject is, then, what I have neither seen, experienced, nor been told, what neither exists nor could conceivably do so. I humbly solicit my readers' incredulity.
>
> Lucian of Samosata, *True History*

As long as humans have imagined anything, we've imagined going into space. As a literary genre, we can date space travel science fiction to long before the Renaissance, all the way back to *True History*, by Lucian of Samosata (AD120–c.180), which begins an entire literary genre loosely called 'imaginary voyages'. *True History* was a second century proto-*Star Wars* saga, with sailing ships propelled into space by violent storms, battles between warring factions and nightmarish Hieronymus Bosch-esque descriptions of alien worlds: the moon was populated by a race of hairless homosexuals, who copulated with each other via openings above the calf muscle. The resulting leg growth was then lanced, the child pulled out, and with mouth open to the wind the life force would blow into them.

Lucian, like Jonathan Swift* in his fantasy travelogue *Gulliver's Travels*, played with the idea of fact and fiction, weaving fantasy and political satire against a celestial backdrop. The universe was no longer distant and immutable, but a place of revelation. A stage where dramas could play out. Disappointingly, like *Star Wars*, Lucian finished with a rather lazy 'To be continued…'. Unlike George Lucas, he never got round to the sequels.

With all science fiction, whether first century, seventeenth century or twenty-fifth, dream worlds and real worlds orbit, collide and cross-pollinate with what we know to be true and what we imagine may become. It's our incomplete knowledge that gives rise to the wildest reaches of the imagination. Imaginative flights of fancy are still our most powerful form of transportation, beyond anything we could engineer.

But of course nothing will ever happen in the real world unless you imagine it first. John Wilkins at least offers some practical advice to the would-be space traveller in his treatise *Mathematical Magick* for those attempting to defy 'gravitas':

* **By spirits, or angels**
* **By the help of fowls**
* **By wings fastened immediately to the body**
* **By a flying chariot.**

The first three options are problematic. Option four – the flying chariot – has provided the more successful basis for modern space travel.

But where will such a machine take us? Where is space exactly?

Left: **Egg from moon goose colony**
© Agnes Meyer-Brandis, VG-Bild Kunst 2017

---

* The great-great-nephew of Francis Godwin.

WHERE DOES SPACE START?

For me space begins near junction 15 of the M6, in between Ashbourne and Leek. I'm at Alton Towers theme park, with Al Worden, the charmingly funny Apollo 15 Command Module pilot, who flew to the moon and back in 1971, earning himself the title of 'world's most isolated human' as a result. Al had remained alone in lunar orbit while the two other crew members were exploring the lunar surface. We're here doing some interviews for World Space Week, as well as being guests of honour for the opening of the park's latest space-themed virtual reality roller coaster, *Galactica*. During the interviews, Al held court with a group of local schoolkids, recounting his extraordinary lunar adventure. There wasn't a flicker of resentment from him, despite having answered the same questions for the last half century: *What's it like on lift off? How do you know which way to go? How do you go to the loo in space?*

As dusk fell, we made our way to the roller coaster for the grand opening and duly christened the new ride. Al, with his NASA flight jacket on, and I were manhandled into the front-row seats of an empty rollercoaster. Restraining bars tightened. Walkie-talkies crackled. Final checks were carried out and we prepared for launch. Fifty years ago, he, along with Commander Dave Scott and Lunar Module pilot Jim Irwin, had been through a very similar procedure in a machine that had orders of magnitude less computing power behind it than the one we were now strapped into. The latest in virtual reality headsets were lowered over our faces like the Apollo spacesuit helmet visors, thrusting us into a computer-generated dream state. The laws of physics, engineering and computer science, woven together with our most profound fantasies. As we rattled on our way, the film in front of our eyes immersed us in a future vision (slightly pixellated) of space travel. Like the mighty Saturn V rocket, designed by Wernher von Braun, the roller coaster car

climbed the arduous gravity well, until we reached the top. All that potential energy was now released as we dropped down the other side in free fall. I'm weightless for a moment, looking at the spherical earth against the blackness of space. A virtual 'overview effect'. Leaving the planet is much easier, safer and cheaper when it's just your imagination that has to do the work. And for that it helps to be on a VR roller coaster at Alton Towers, while sitting next to an Apollo astronaut.

The good news for you, wherever you are in the world, is that space isn't very far away. A hundred kilometres to be exact. An easy commute in a straight line. That's the distance from where I'm sitting now (Rosebery Avenue in London) to Portsmouth on the south coast or Northampton. A lot closer than Alton Towers.* A little past Oxford or Cambridge. A train ticket to Portsmouth will cost me £10 and take an hour or so. Seven hours on my bike, according to Google Maps.

There is no physical boundary or natural shoreline marking where space begins, just a gradual thinning of the atmosphere as the air molecules become spaced further apart. The Fédération Aéronautique Internationale (FAI) based in Lausanne, Switzerland, is the official governing body that acknowledges and mediates aerospace records, including human space flight. They are the organization who patrol this imaginary boundary and it is them you will need to convince that you've left the planet and entered space.

There was controversy surrounding Yuri Gagarin's first flight, because officials submitting the documentation of the event had not mentioned that on his return to earth he'd ejected from his spacecraft (Vostok 1) at 20,000 feet. The FAI had stipulated at the time that you had to land with your spacecraft for the record to stand, but – reflecting the spirit of the law rather than the letter – sensibly, in light of the historic achievement of Gagarin's flight, this rule was amended.

* 203 km as the goose flies.

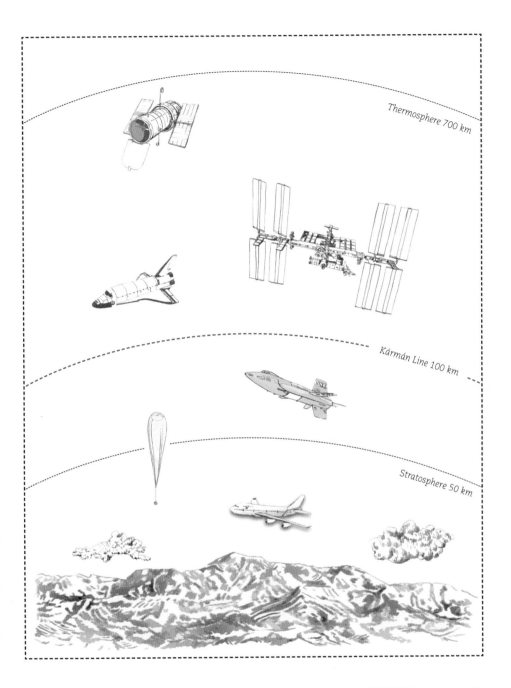

Thermosphere 700 km

Kármán Line 100 km

Stratosphere 50 km

CROSSING THE LINE

The Kármán Line is an imaginary line 100 kilometres above sea level named after the Hungarian physicist and aerospace engineer Theodore von Kármán. In an informal set of discussions in the mid-1950s with scientists and engineers from the FAI and the, confusingly for us, IAF (International Astronautical Federation), von Kármán established the demarcation line between *aeronautics* ('sailing the air') in the atmosphere, where aircraft need aerodynamic forces to work, and *astronautics* ('sailing through the stars') beyond the atmosphere, where aerodynamic flight is no longer feasible.

As the space race heated up in the early 1960s, aircraft such as the hypersonic rocket-powered X-15 began blurring the line between aircraft and spacecraft, so an official space boundary became important. But this demarcation had international differences at the time. A handful of X-15 pilots, including Bill Dana, John B. McKay and Joseph A. Walker, were awarded 'astronaut wings' for 50-mile (80-km) flights. Von Kármán had very sensibly suggested 100 kilometres to be the internationally-recognized boundary line, being a nice round number and easy to remember, and that eventually stuck. Crossing that line now means you get your astronaut wings, which is ironic since the last thing that will help you in space is wings, but excitingly it also means a badge and/or lapel pin along with bragging rights. Who you are and where you're from will determine what kind of badge you'll get. Unless you're a member of the US armed forces, or flying as a NASA civilian, you'd most likely be going for a Federal Aviation Administration (FAA) civilian astronaut badge.

So far, only SpaceShipOne test pilots Mike Melvill and Brian Binnie have this civilian astronaut badge, so you'd be in exclusive company. If you're planning on becoming a 'Bransonaut' with Virgin Galactic they have designed their own pin. If you're British born, the British Interplanetary Society will award you one of their silver rocket pins with a Union Jack. Private astronaut Richard Garriott lost his and asked for another one. I had the honour of awarding this replacement to him over a pint with some friends in the Cittie of Yorke pub in London's High Holborn.

GRAVITY WELLS

If space is so close, within walking distance, why is going 'uphill' so tricky? In a word: gravity.

Isaac Newton first described mathematically in his work *Philosophiae Naturalis Principia Mathematica* why your ballpoint pen won't write upside down and how objects like apples falling toward the centre of the earth are ruled by the same force keeping the planetary bodies in orbit. Gravity, of course, keeps us planted on the ground, keeps the tides moving (thanks to the moon) and keeps the atmosphere in place. All very useful until you want to go somewhere else. We tend to think of gravity as being very strong, holding large structures like solar systems and galaxies together. But the fact that you, weighing 70 kg or thereabouts, can jump in the air under your own muscle power and separate yourself from something as massive as a planet weighing 5.9721986 x 10^{24} kg, means the opposite is true. Giant stars may collapse under gravity, but a cheap airport-bought novelty magnet sticks happily to the fridge door despite the mass of the entire earth trying to dislodge it. This is all summed up neatly for you in Newton's famous gravity equation:

$$F = \frac{GM_1M_2}{r^2}$$

I'm momentarily free from the effect of gravity on the apexes of the Alton Towers roller coaster with Al Worden, for exactly the same reason that astronauts float about on the International Space Station. Tim Peake

Above: **British Interplanetary Society astronaut pin**

is floating, not because there's an absence of gravity, but because he's constantly falling around the earth. If he wasn't moving, he'd be walking around normally.

BREAKING FREE
The secret to getting into orbit and staying there is to travel fast enough, and not to slow down. Around 17,000 mph will do it. The secret to breaking free of the earth altogether, is to go even faster: say 25,000 mph. We call this earth's *escape velocity*.

Let's say you throw a tennis ball or an apple as hard as you can – how strong you are will determine how fast, and therefore how far, it will travel. You'll notice as its speed drops, the path of the ball curves (a parabola) towards the ground due to objects – planets and tennis balls – being attracted to each

other by gravity. Because the earth is curved,[*] if you can throw the ball fast enough, say at over 17,000 mph, its curved path is big enough to miss the earth altogether and go on missing it. If your parabola matches the curve of the earth, it will keep falling round and round the planet. A satellite or space station will stay on its trajectory around the earth as long as you want it to because there's nothing to slow it down, like air resistance.

The smaller your planet, the less gravity there is and therefore the easier it is to leave. Getting off the moon takes a lot less energy than getting out of the deeper *gravity well* of the earth. Even more massive objects, like stars or giant planets, have even deeper gravity wells. Black holes, which aren't holes at all but extremely dense collapsed neutron stars, are the deepest of them all.

---

[*] Due to gravity. If the radius of an object is greater than the so-called 'potato radius' of about 200 km, there will be sufficient gravity pulling towards its centre to shape it into a sphere.

FALLING AROUND THE EARTH

The key to thinking about gravity and orbits and getting to and from the moon is in the word *falling*. The moon is falling around the earth. As is the International Space Station (ISS). If the ISS slowed down by firing its rocket boosters in the opposite direction of travel, it would slow down and fall back to earth just like your tennis ball.

At the time of the Jacobean space programme, our understanding of gravity was still in a muddle. It was reasonably assumed that heavier objects, when dropped, accelerate to the ground faster than lighter ones. Galileo Galilei in 1589 famously (apocryphally) dropped two different weights off the Leaning Tower of Pisa to show that in fact they fall at the same speed. You can try this: if you drop, say, a hammer and a feather then it seems reasonable to assume that the hammer will hit the ground first. You'd be right, but only because of their *shapes* – air resistance slows the feather more. If you go to the moon (where there is no air), and specifically to the Apennine mountain range at the Apollo 15 landing site, you will see a geological hammer and a falcon feather lying in the lunar dust. If you pick them up, hold them at arm's length and drop them, you'll notice they indeed land at the same time, which was confirmed by Commander Dave Scott when the experiment was filmed on the moon in 1971: 'How about that? Mr Galileo was correct in his findings…'

It wasn't until Newton's description of gravity in *Principia* in 1687 that the basics were cleared up. Isaac Newton in your flying chariot's driving seat will get you into orbit and to the moon and back. Albert Einstein went even further in describing what gravity actually was and how it worked. In his theory of general relativity[*] he realized that massive objects like stars and planets warp the fabric of space-time itself, like bowling balls on stretched rubber sheets.

You can demonstrate the time-warping effect of the planet's gravity field yourself, which I did once on the BBC science programme *Bang Goes the Theory*. All you need are a couple of synchronized, highly accurate caesium beam atomic clocks, a long-range aircraft and a pilot to fly you and one of the clocks around the world. When you reunite the clocks, you'll notice they're no longer synchronized.[**] A fun experiment to try, although atomic clocks and airliners are generally harder to get hold of than hammers and feathers.

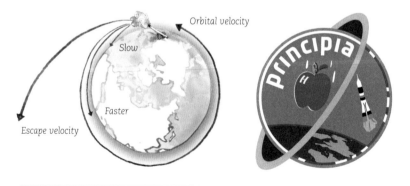

Orbital velocity

Slow

Faster

Escape velocity

[*]
$$R_{\mu\nu} - \frac{1}{2} R\, g_{\mu\nu} + \Lambda\, g_{\mu\nu} = \frac{8\pi G}{c^4}\, T_{\mu\nu}$$

[**] See Hafele-Keating experiment, 1971.

Left: **Newton's cannon**
Above: **Tim Peake's mission patch**
Right: **SuitSat. A retired spacesuit set adrift from the ISS**

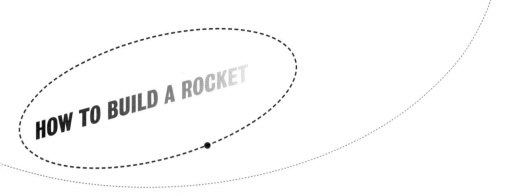

HOW TO BUILD A ROCKET

'When you want to build a ship, do not begin by gathering wood, cutting boards, and distributing work, but rather awaken within men the desire for the vast and endless sea.'
Attributed to Antoine de Saint-Exupéry

MAKE YOUR OWN ROCKET

1 Find a little plastic 35mm camera film pot. Not so easy to find these days but if you look in that drawer in the hallway, right at the back there'll be one. Make sure the lid snaps shut on it. A good clean tight seal is what we're looking for.

2 Get an Alka-Seltzer from the bathroom cabinet, break it up a bit and if it's one of the big ones put about a third in.

3 Speed is of the essence here: fill the pot with water up to a fraction under halfway. You'll hear the tablet fizz as it reacts. Quickly snap the lid on (making sure you hear the snap), give it a quick shake and then put it on a table upside down.

4 Count down from ten. If you're lucky, the moment you say LIFT-OFF the pot, minus the round lid, will launch into the air, hitting the ceiling or travelling up a good twenty feet if you're outside.

5 Seek approval from your audience, and a cloth to wipe up the mess.

SEALED PRESSURIZED INK CARTRIDGE

Ultra-hard Tungsten Carbide Ball

Stainless steel precision socket prevents leaking yet delivers uniform ink flow

Thixotropic ink in a hermetically sealed, pressurized reservoir writes three times longer than an ordinary ballpoint

Sliding float separates ink from pressurized gas

ANTI-GRAVITY TECHNOLOGY

Gas Plug

it looks like a machine from another era. Something designed in the 1950s, rather than a twenty-first-century space travel vehicle. That's because it is. Everything at the Cosmodrome is frozen in a Cold War time capsule – like the fading photographs still displayed on the walls, bleached by sunlight, of the doomed Soviet Buran space shuttle and the giant N1 moon rocket that was put together in this very hangar but never made it off the launch pad.

The Soyuz launch vehicle, like the famous Saturn V moon rocket, is a series of rockets on top of rockets, all stuck together like Lego forming a stack called the *stages*. We can imagine its three stages working a bit like going through the gears of a car as it accelerates. The rocket equivalent of a car's gear change is a spent stage falling off and the next one on top igniting. Inside the top stage of the Soyuz launch vehicle is the Soyuz spacecraft itself, made up of three modules: *orbital* (which docks with the space station and has a loo in it), the *descent* module (the bit where the astronauts sit), and the *service* module (the bit with all the gubbins). It's worth remembering that almost nothing of the Soyuz launch vehicle and spacecraft survives the journey to space and back: the rocket stages fall back to earth downrange, and the orbital and service modules of the spacecraft burn up on re-entry. The only bit that lasts to the bitter end is the tiny descent module that parachutes the astronauts back down to the ground. You can see the descent module and the huge parachute that brought Tim Peake and his crew back to earth in the Science Museum in London.

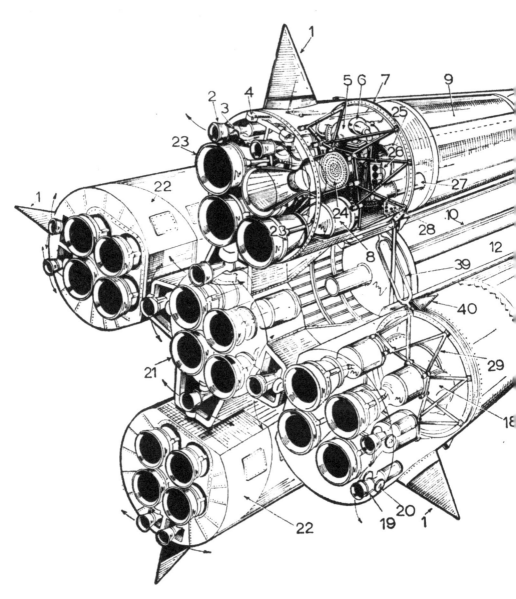

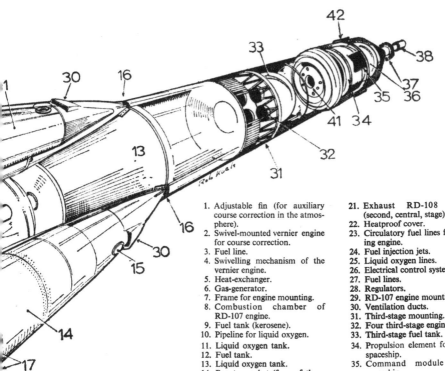

1. Adjustable fin (for auxiliary course correction in the atmosphere).
2. Swivel-mounted vernier engine for course correction.
3. Fuel line.
4. Swivelling mechanism of the vernier engine.
5. Heat-exchanger.
6. Gas-generator.
7. Frame for engine mounting.
8. Combustion chamber of RD-107 engine.
9. Fuel tank (kerosene).
10. Pipeline for liquid oxygen.
11. Liquid oxygen tank.
12. Fuel tank.
13. Liquid oxygen tank.
14. Booster rocket (four of these, which together form the first stage).
15. Tank aperture.
16. Separation point between booster rockets and second stage.
17. Booster aerial.
18. Turbopump.
19. Vernier engine.
20. Swivelling mechanism of vernier engine.
21. Exhaust RD-108 engine (second, central, stage).
22. Heatproof cover.
23. Circulatory fuel lines for cooling engine.
24. Fuel injection jets.
25. Liquid oxygen lines.
26. Electrical control system.
27. Fuel lines.
28. Regulators.
29. RD-107 engine mounting.
30. Ventilation ducts.
31. Third-stage mounting.
32. Four third-stage engines.
33. Third-stage fuel tank.
34. Propulsion element for Soyuz spaceship.
35. Command module Soyuz spaceship.
36. Cabin space of Soyuz (can also be used as airlock).
37. Escape system.
38. Separation rockets for escape system.
39. Cruciform booster link.
40. Streamlined cover.
41. Rear of Soyuz with engines for course correction and braking.
42. Stabilisation flaps for escape system.

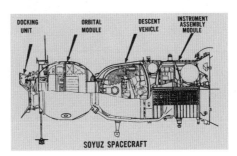

DOCKING UNIT   ORBITAL MODULE   DESCENT VEHICLE   INSTRUMENT ASSEMBLY MODULE

SOYUZ SPACECRAFT

КОРОЛЁВ* **CROSS**

The most beautifully simple parts of the whole machine, in my opinion, are the four, plain-looking brackets attached to the central core block stage, which the ends of the four cone-shaped strap-on boosters plug into. The pointy end of the booster slips inside the bracket, a bit like a ball and socket joint. After the rocket is pulled out horizontally to the launch pad and slowly raised into its vertical position, it sits in place on top of the boosters supported by these load-bearing brackets. Pivoting gantry arms with large yellow barrel-like counterweights are pulled down creating a support ring, and kept in place under the weight of the rocket sitting on them like a golf ball on a tee. The support structure also doubles as an altar when the rocket is blessed with holy water.

When the boosters (first stage) ignite, followed moments later by the central core (second stage), the rocket becomes weightless and begins to rise. As it does, the supporting gantry arms fall back under the weight of their concrete counterweights and the rocket is released like an insect escaping from the petals of an opening flower. The four boosters push the rocket upwards against the brackets until they run out of steam, whereby a retaining strap at the base of the boosters is severed using pyrotechnics, and they simply fall out of the socket under their own weight, falling back to earth. Newton is in control at every stage of the process. A little venting liquid oxygen at the top of each booster provides a nudge to make sure they're pushed away from the central core.

If it's a clear day, you can see the four boosters tumble away in a crucifix shape, known as the КОРОЛЁВ cross. That's the priest's favourite bit. After a few minutes the third stage ignites, pushing away the spent central core, which also falls to the ground, sending the spacecraft itself into orbit. Even

* Korolev.

though we're fighting the force of gravity, that force also becomes our friend as the rocket sheds weight as it climbs. No launch towers to bump into. It's simple and elegant – reliable, dependable, replaceable. When you're sending people into space on top of a bomb, you don't want cutting-edge. You don't want experimental. You want a system with as few parts as possible that work every time. Like a pencil.

Down range of the Cosmodrome, in the wide open tundra, spent boosters, fuel tanks, escape towers and nose-cone shrouds rain down on the ground like breadcrumb offerings thrown to the birds. Locals were known to collect them for scrap metal, roofing and kids' sledges, or to sell on to unscrupulous dealers. One of the first stage booster's internal cylindrical tanks happens to be shaped like a type of local cooking pot and is said to be salvaged by rocket hunters and used to cook the traditional Khazac lamb stew.

Other rockets have come and gone. But this Russian R-7 family of rockets has outlasted them all. Since the Space Shuttle made its final flight in 2011, there's been no crewed American space vehicle. The three seats on the Soyuz are divvied up between all nationalities who want to go to the ISS, as well as the small handful of pioneers who've personally forked out upwards of $20 million for the privilege. The only other crewed space rocket that can get you into space at the time of writing is the Chinese 'Long March' rocket to their own Tiangong space station.* Fine if you're Chinese, but for the rest of us, at least for now, Baikonur is the only bus stop to space, at the same launch pad from which Sputnik, Laika the dog, Gagarin, Valentina Tereshkova, Helen Sharman and now Tim Peake launched.

This simple rocket system has launched everyone and everything except the one thing it was originally designed for – a nuclear warhead.

---

* 'Heavenly Palace'. The Chinese have the best space names.

THE FIVE FOREFATHERS OF ROCKET SCIENCE

Scientists, engineers and dreamers fuelled not just by chemical reactions but by an even more potent blend of politics, philosophy, art and imagination.

1. THE RUSSIAN MYSTIC VISIONARY OF ROCKETRY

'The Earth is the cradle of humanity, but mankind cannot stay in the cradle forever.'
Konstantin Tsiolkovsky

The grandfather of Soviet rocketry was Konstantin Tsiolkovsky (1857–1935), who came from a poor family and lost much of his hearing due to an illness as a child. He was self-schooled in maths and lived much of his life in a log cabin. Tsiolkovsky was deeply influenced by the transhumanist spiritual philosophy known as Russian *cosmism*, believing that the future of the human species must be elevated through science: in particular, to transcend the bonds of earth and journey into the cosmos. Many of these ideas were expressed in his drawings.

He was fascinated by the writings of science-fiction author Jules Verne and worked out that Verne was at least theoretically correct in *From the Earth to the Moon*, about the idea of firing an artillery shell-type craft into space, but that the energy unleashed at the moment of ignition would crush the occupants. He realized that a reaction engine was the way to go, and most importantly it would work in a vacuum.

He started doing the calculations to work it all out and eventually came up with his famous rocket equation[*] in 1903, the year of the Wright Brothers' first powered flight.

Fall in love with it. If you want to go to space it's the only way to go.

$$\Delta v = v_e \ln \frac{m_0}{m_f}$$

'First, inevitably, the idea, the fantasy, the fairy tale. Then, scientific calculation. Ultimately, fulfilment crowns the dream.'
Konstantin Tsiolkovsky

---

[*] Basically, the more fuel you have, and the lighter the rest of the rocket, the better.

2. THE AMERICAN BACKYARD ROCKET ENGINEER WHO LIKED TO CLIMB TREES

Robert H. Goddard (1882–1945), the father of American rocketry, was less of a mystic and more a man of action. A teenage Goddard was asked by his father to trim the cherry tree in the garden, which is where he had his epiphany – a vision of space flight and the machines that could make it happen. If Tsiolkovsky was the visionary and autodidact, Goddard was a practical experimenter, with a Masters degree and a PhD from Clark University in physics. Even when he was asleep, Goddard was working. On 8 August 1915, dreaming of standing on the moon, he noted in his diary:

> Set off red fire at a pre-arranged time… so all [on earth] can see it. Was cold… took photos of Earth with small Kodak while there – two for stereoscopes… Not enough oxygen when I opened my helmet to see if so… light was rather dim.

There was a section at the end of his definitive 1919 work, *A Method of Reaching Extreme Altitudes*, called *Calculation of Minimum Mass Required to Raise One Pound to an 'Infinite' Altitude*. It was the word 'infinite' that got everyone excited, and was picked up by the popular press – something that Goddard, a private man, hated. In 1930 he cautiously outlined how such a rocket would work. He mentioned a potential solution to the problem of refuelling, by using chemicals on the moon or Mars, something that is still talked about to this day:

> The great rocket, with a compartment large enough to hold a passenger comfortably with all of the equipment necessary for his task, will not need to carry propellants for the return trip. The chemical elements necessary for refuelling exist on the moon to the most careful astronomical observations, and it will be possible for the passenger to take off without any outside assistance for the return journey.
> *Modern Mechanix*, January 1930

16 March 1926 is a date that should be as familiar to us as that of the Wright brothers' first flight. On 'Aunt' Effie Ward's farm in Auburn, Massachusetts, Goddard launched his first rocket using liquid oxygen and petrol. Called the 'Goddard 1' or 'Nell', it rose 12.5 metres or 41 feet, a journey lasting 2.5 seconds. The event was being recorded for posterity by his wife, Esther. Unluckily for posterity (and Esther), the wind-up camera was being re-wound at the crucial moment.

We have much to thank Goddard for. He was the first to use vanes for aerodynamics, to propose (independently of Tsiolkovsky) a multi-stage rocket system, and to launch a payload of scientific instruments (camera, barometer, thermometer). He continued his experiments over the years, eventually launching rockets up to 9000 feet in Roswell, New Mexico. A location that was to become infamous for a different sort of dreamer…

3. THE CHIEF DESIGNER

For many years the architect of the Russian space programme was a mystery: known only in the West as the 'Chief Designer', his identity and work were shrouded behind an iron curtain of secrecy. It was only after his death that he became widely known as Sergei Pavlovich Korolev (1907–1966).

As a young man he studied aeronautical engineering where his interest in rocketry and spaceflight developed, and soon he was experimenting with liquid-fuelled rockets. In 1938, during Stalin's great purge, he was denounced by a fellow engineer and tortured in the Lubyanka prison in Moscow, before being sentenced to ten years in a Soviet Gulag. There he endured appalling conditions, freezing climate, scurvy and losing all his teeth. Instead of serving the full sentence Korolev was moved to a prison for intellectuals and academics, and was eventually released in 1944.

After the war he set to work on a national missile programme, whose aim would be to create the first intercontinental ballistic missile (ICBM) that could carry a nuclear warhead over the North Pole to America. But Korolev understood the national prestige of using such a rocket to conquer the new frontier of space. In October 1957, he oversaw the world's first artificial satellite, Sputnik 1, launched by the R-7 missile with its four tapered strap-on boosters. The polished metal sphere was hurriedly put together as a glorious statement for the International Geophysical Year*, falling around the earth, broadcasting its *beep, beep, beep* and changing the world for ever. The space race had started. Korolev was offered the Nobel Prize twice for his achievements but had to turn them down to preserve his anonymity. The engineers' work forever tethered to the march of politics, culture and conflict: how important these issues on the ground which melt away when seen from orbit.

4. HERMANN OBERTH, A WOMAN IN THE 'MOONE' AND THE NAZI FLYING SAUCERS

> '"NEVER" does not exist for the human mind... only "Not yet".'
> *Frau im Mond*, Fritz Lang

Perhaps no science fiction film in the history of cinema has been quite so prophetic as Fritz Lang's 1929 silent moon voyage film *Frau im Mond*. If you're heading above the atmosphere for whatever reason, put it on your list of films to watch before you go.

That year the waves of the Wall Street Crash in America were slamming through the

German economy. At the film's heart is the central premise that the moon might be rich in gold. The plot and stylized performances resemble a German expressionist version of the James Bond film *Moonraker*, and follows the fortunes and adventures of an unlikely band of lunar adventurers. The themes of the film, from the quasi-spiritual, to brash American populism, to dictatorial fascism, reflect the real protagonists that would dominate twentieth-century rocket science and the politics of the day.

But the most interesting aspects for our purposes are the science and rocket engineering behind this fictional moon voyage. The parallels between the journey to the moon in the film and the real history of space travel are striking. The first rocket we see is a reconnaissance probe, with on-board

---

* The International Geophysical Year (IGY) was an 18-month international scientific endeavour to explore and better understand the workings of our planet.

cameras and a set of scientific instruments to gather information about the moon. It looks remarkably like the famous V-2 – Hitler's weapon – that was used fifteen years later, but this one is landing fully upright on its fins much like Elon Musk's SpaceX rocket Falcon 9. The giant 'Spaceship Friede', the ship that will carry our unlikely heroes to the moon, is first seen emerging from a giant hangar, vertically stacked on a flat transporter to be moved to the launch pad, just like the real Saturn V and Space Shuttle.

With the rocket on the launch pad we see the first-ever backwards rocket countdown, beloved of Hollywood space films ever since. The final 3, 2, 1… ends with a dramatic single card: JETZT.* The countdown is an invention not of rocket science but of cinema, intended to ramp up the tension for the audience, and something NASA has used throughout its history for exactly the same purpose. Just like real astronauts, the crew in the film are launched lying face up on spring-loaded bunks, allowing them to withstand more of the G force. We see liquid fuel being used, and rocket stage separation, and when we arrive at the moon, only the crew compartment lands on the surface. On board the ship, the effect of weightlessness is considered and an engineering solution devised in the form of leather floor straps. We see a spacesuit like a Victorian diving suit.

Most significant is the view from the spaceship's window. Here we see the earth as a sphere from space. The crew's reaction to the view is the very first cinematic example

Above: **Gerda Maurus stars in *Frau Im Mond***

* NOW

of humans experiencing the 'overview effect' – the profound and emotional sense of context that astronauts get looking back at the earth. Our film's protagonists, good and evil, young and old, all stare back at the earth for that single moment in unity – just as we've always done in our fantasies, and as humanity did when we eventually saw these images for real.

Where did the detailed knowledge for the film come from, and why was it so accurate in foreseeing future rocket design and human space flight? Hermann Oberth (1894–1989), the father of German rocketry, was brought on as a technical adviser for the film. As a boy he memorized all the Jules Verne novels, and like Tsiolkovsky considered the implications of

Above: **The Frau im Mond rocket Friede**
Right: **Space Shuttle Endeavour**

firing a projectile from the earth to the moon. Oberth's work inspired a generation of rocket enthusiasts and, unlike Goddard, he was open with his ideas, engaging with the popular culture of the day. As a result, German rocketry flourished and spawned some unusual offshoots, most infamously Gerhard Zucker's failed Western Isles 'rocket mail' from Harris to Scarp – a dramatic forerunner of Amazon's proposed drone-package delivery idea, but which only resulted in a large bang and some badly singed letters.

Oberth later became an advocate of flying saucers – a craze sweeping popular media from the late 1940s. If you reach for your copy of *Flying Saucer Review*, Volume 1,

Above: **The *Frau im Mond* film set**

Number 2, from 1955 (the one with the red cover) you'll see his gloriously titled 'They Come from Outer Space' in which he outlines his thoughts on modern UFO phenomena, suggesting that at least a small percentage of UFOs were extraterrestrial in origin, although he says earthly possible explanations such as flocks of wild geese[*] shouldn't be ruled out.

But it's for his work at Peenemünde, the secret missile base on the northern German coast, on the infamous V-2 rocket, the world's first ballistic missile, that Oberth is best known, and for his association with the most famous rocket scientist of them all: the man whose work was inspired by Oberth, who would take all these ideas and build the first rocket to break free of gravity and our imagination and fly us to the moon.

[*] Gansa?

5. DR SPACE AND THE VENGEANCE WEAPON

Baron Wernher Magnus Maximilian Freiherr von Braun (1912–1977) was a rocket scientist with unique charisma and PR skills. He understood how political aims could be harnessed and exploited to further his own technological fantasies, and through his work and unique personality he became a worldwide celebrity.

From a young age von Braun built rocket-powered go-carts in which he'd terrorize his local town. He was fascinated by Oberth's theories and took it upon himself to learn maths and physics so he could understand them better. As a teenager he joined the enthusiastic amateur rocket society Verein für Raumschiffahrt, of which Oberth was a founder member. After the First World War, Germany was heavily restricted from rearming. But with the rise of the Nazis, there was growing military interest in the possibilities of rockets as weapons of war. Von Braun and his team were commissioned to work on new rocket projects.

It was von Braun's mother who suggested Peenemünde – a remote area on the German Baltic coast – as a suitable area for a secret rocket research centre. Established in 1937, Peenemünde quickly became a world-class facility. It was here that the first 'Aggregat' family of ballistic missiles were developed. The most infamous of these new rockets, with a range of 200 miles, far beyond the reach of contemporary artillery, and the ability to carry a one-ton warhead that would explode on impact, became known as the V-2 *Vergeltungswaffe* or 'vengeance weapon' that would rain down on London and Antwerp during the death throes of the war.

On 3 October 1942, an experimental V-2 rocket christened *Frau im Mond* after Lang's film was successfully launched.* This was the day, according to General Walter Dornberger, the military head of the V-2 programme, that the spaceship was born: 'This third day of October, 1942, is the first of a new era in transportation, that of space travel...'

At the end of the war, von Braun and many of the Peenemünde engineers were spirited away to the United States to work in the new American space industry. The work von Braun would do in America would see the launch of his rockets carrying the first American satellite, the Explorer 1, culminating eventually in the mighty Saturn V, which would take us to the moon.

Von Braun was able to ride the wave of an emergent popular culture in 1950s America. A population seduced with stories of rocketry and space travel. Walt Disney brought von Braun's space vision to the people, producing a series of specials featuring him to promote the newly opened Disneyland. In these TV films, watched by millions, we see von Braun in front of the camera delivering his vision of things to come, seducing the public hungry for a bright, bold technological future beyond the earth.

Of course with the von Braun story comes the baggage of his political past. In Germany he had been an SS *Sturmbannführer* (Major), a fact that he always defended as being simply a means to an end, but a fact, nonetheless, overlooked by those focused on the new Cold War and beating the USSR to the moon. His opportunism and conflicted reputation is elegantly summarized by the great American songwriter Tom Lehrer in 'Wernher von Braun', reprinted here for you to play and sing at home with his generous and good-humoured permission.

Opposite: **V-2 rocket with a *Frau im Mond* logo**

---

* Ironic, given that Friede means 'Peace'.

Wernher von Braun

Words and Music by
Tom Lehrer

Gently

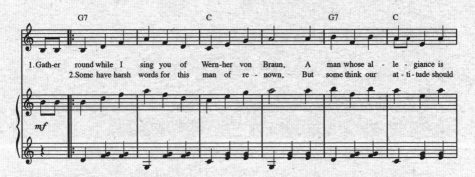

1. Gath-er round while I sing you of Wern-her von Braun, A man whose al - le - giance is
2. Some have harsh words for this man of re - nown, But some think our at - ti - tude should

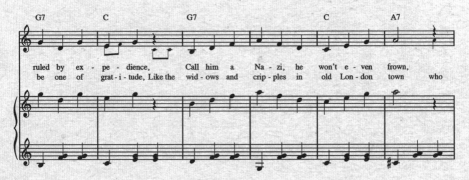

ruled by ex - pe - dience, Call him a Na - zi, he won't e - ven frown,
be one of grat - i - tude, Like the wid - ows and crip - ples in old Lon - don town who

"Na - zi, Shma - zi," says Wern - her von Braun. Don't say that he's hyp - o -
owe their large pen - sions to Wern - her von Braun. You too may be a big

crit - i - cal, _____ Say rath - er that he's a - po - lit - i - cal, _____ "Once the
he - ro, _____ Once you've learned to count back - wards to ze - ro, _____ "In

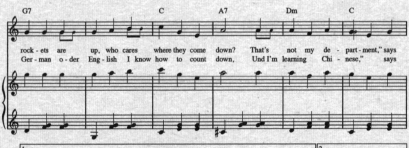

rock - ets are up, who cares where they come down? That's not my de - part - ment," says
Ger - man o - der Eng - lish I know how to count down, Und I'm learning Chi - nese," says

Wern - her von Braun. Wern - her von Braun.

YOUR FUTURE RIDE TO SPACE

Plans are afoot to expand our capabilities for taking more people into space, assisted by entrepreneurship, international cooperation, government and private partnerships, revelations in material science and digital technology. And of course imagination. If you start saving now, these are some of the vehicles and spacecraft that will be taking you off the planet in the next decade or so.

Opposite: **Apollo 11 launch**
This page: **Space Shuttle** *Endeavour*

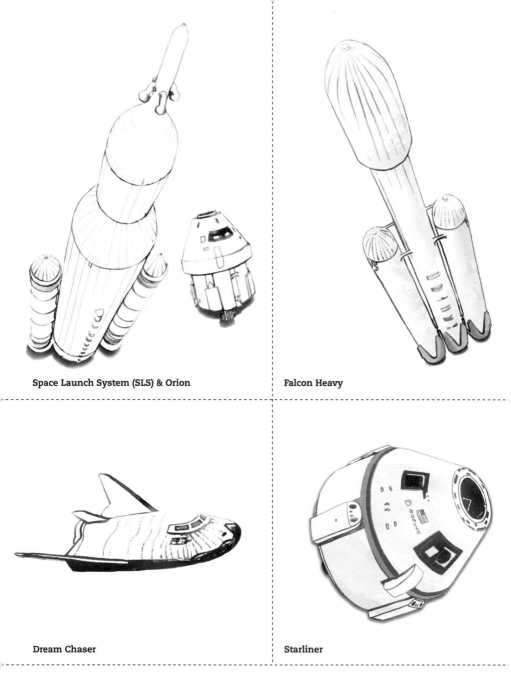

Space Launch System (SLS) & Orion

Falcon Heavy

Dream Chaser

Starliner

NASA – SPACE LAUNCH SYSTEM (SLS)/ORION

The SLS will be NASA's first launch vehicle for taking people into space since the Space Shuttle. It is one of the many components needed to get us to Mars and is designed to be adaptable to the changing demands and destinations of deep space exploration. If it's raw power you want this is the rocket for you: it will be the most powerful ever built.

NASA's Constellation rocket system programme was scrapped by the Obama administration, and who knows now where we're heading politically. However, this new rocket design is going ahead and into commission, and we're told we could be travelling to Mars by the mid-2030s. The reality is that large-scale projects like Apollo or a mission to Mars require long-term planning, much longer than self-serving political cycles can facilitate. The current mood seems to be pointing towards a NASA focused on the big showroom missions like the moon and Mars, leaving the ferrying of astronauts to low earth orbit to commercial partners like SpaceX and Boeing.

You can have it in any colour you like so long as its orange, which is the colour of the liquid hydrogen and LOX tank insulation foam that covers the body of the rocket. Painting it white as originally planned would have added an extra 1000 lbs (450 kg) and sending that amount of weight into space would cost upwards of $5 million.

For our purposes, the important bit is where you're going to sit. For the SLS this is NASA's Orion spacecraft built by Lockheed Martin, a 'multiple purpose crew vehicle' made up of the crew module, a European-built service module which carries your solar panels, propellants, oxygen, etc, a 'Launch Abort System', and the stage adapter that keeps it all safely on top of the SLS rocket. The crew module is where you'll be sitting. It resembles the Apollo capsules from the 1960s but it's roomier – designed to carry four people rather than three. Its design principles though are the same: a flat, heat-shield-first trajectory for ballistic re-entry* through the atmosphere, which will be slowed down by parachutes and a splash-down landing and recovery at sea. The spacecraft has already been tested in space in 2014, going further than any crew-rated vehicle since 1972, and it works.

Once you're inside though that's it. It's not specifically suited for lengthy Extra Vehicular Activities (spacewalks), or indeed landing anywhere. It's perfect for a romantic flight around the moon, helping to build another spacecraft in circumlunar space to take you to Mars, or checking out some valuable asteroids. Quite small windows though, so if it's the overview effect you're after you might want to look elsewhere.

SPACEX – FALCON/DRAGON

Elon Musk's SpaceX isn't just a company that wants to deliver astronauts and cargo locally. Musk's vision is nothing less than colonizing Mars and making humans a multi-planet species. These days the dream of space travel is being realized by the super-rich, themselves inspired by the legacy of Apollo and the science fiction of the 1970s and 1980s. Musk's Falcon 9 rocket has delivered satellites into orbit, but the fun bit is that the used first stage lands upright back on earth just like rockets are supposed to. There are two ocean barges currently serving as landing platforms named 'Just Read the Instructions' and 'Of Course I Still Love You' after the starships in Iain M. Banks's novel *The Player of Games*.

The Falcon 9 'Heavy' (now just called the Falcon Heavy) is the beefed-up version for larger loads. It's essentially a Falcon 9 with two extra liquid-fuelled boosters strapped to the side, designed with heavy payloads and, like the SLS, Mars in mind.

---

* A ballistic re-entry is controlled only by gravity and atmospheric drag – fast, hot and with very high g-force – with hardly any aerodynamic lift from the shape of the spacecraft to soften the ride.

The SpaceX spacecraft that the Falcons will lift is called Dragon, of which there are several variants:

- **Dragon** – this is currently delivering supplies to the ISS, making it the first private spacecraft to do so.

- **Crew Dragon** – Dragon was always designed to eventually carry a crew of seven to the ISS and is scheduled to do so in the coming years. It's a ballistic capsule like Orion, but rather than using a parachute on re-entry, uses eight SuperDraco retro-engines clustered around the side of the spacecraft, which also double as abort thrusters. The spacecraft has four windows.

- **DragonLab** – an orbital microgravity laboratory. Good for doing science. Not so good for exploring the cosmos.

- **Red Dragon** – (now cancelled). An unmanned Dragon capsule that was to be used as a low-cost solution to land on Mars and return home with samples.

SIERRA NEVADA CORPORATION – DREAM CHASER

The Dream Chaser is a mini seven-seater reusable space utility vehicle (SUV) Space Shuttle-type craft – but about a quarter of the size – with an aerodynamic 'lifting-body' (the airframe itself provides the lift) designed to get crew and/or cargo to low earth orbit destinations like the ISS. Rather than a ballistic cone-shaped capsule like Orion or the Starliner (see below), this will glide back to earth like an aircraft on a conventional runway, a design feature that has obvious benefits. It's small enough to be carried on an Atlas V rocket, the commercial launch system developed by Lockheed Martin and Boeing, which has launched satellites and NASA orbiters and probes including the Mars Reconnaissance Orbiter, New Horizons and Juno.

The idea for space planes isn't new. Before the Shuttle the US Air Force's Boeing X-20 Dyna-Soar from the 1950s was very similar in concept to the Dream Chaser: this got to the design and planning stage, including the selection of a young X-15 pilot, Neil Armstrong, but it never flew. Spaceplane development continues in other ventures, such as the British Skylon single-stage autonomous orbital spaceplane with its revolutionary jet/rocket hybrid Reaction Engines, but sadly for us this has no seats.

BOEING – STARLINER

Boeing know a thing or two about building aircraft and spaceships, being a prime contractor for NASA from the 1960s. The Boeing CST-100 Starliner is designed as a seven-seater ferry to get astronauts and you to the ISS and back. Similar to the Orion capsule, it will initially launch on an Atlas V rocket. Like Boeing's 787 Dreamliner aircraft, which you might have already flown in, the Starliner promises to be the very essence of space-chic – with mood lighting, wireless Internet, tablet controls and a brand-new innovative NASA Docking System (NDS). It's exactly what a twenty-first century spaceship should look like inside. The good news is one of these seats will be available for tourists. The bad news is that you can't afford it.

Opposite: **NASA's Space Launch System (artist's concept)**

CAN I TAKE MY DOG?

'All the universe is full of the lives of perfect creatures…'
Konstantin Tsiolkovsky

Yes you can. Wherever we have dared or dreamed to venture, animals have been by our side. Domingo Gonsales's moon geese pull him across space like Antarctic explorer Ernest Shackleton's dogs across the *terra incognita*. In Jules Verne's prophetic *From the Earth to the Moon*, the inclusion of animals was a given:

> Ardan wished to convey a number of animals of different sorts (not indeed a pair of every known species), as he could not see the necessity of acclimatizing serpents, tigers, alligators, or any other noxious beasts in the moon… 'our projectile-vehicle is no Noah's ark, from which it differs both in dimensions and object. Let us confine ourselves to possibilities'…

Whatever your preferred companion, dog, cat, rat, monkey, chimpanzee, spider, newt, fish, jellyfish, frog, tadpole, mouse, stick insect, more mice, guinea pig, Madagascan hissing cockroach, there's a good chance that if it's small enough to fit into the nose cone of a rocket, then it's been into space. As you would expect some species have fared better than others. Some are remembered as heroes; others should never have been there at all; most have been forgotten or relegated to footnotes of the space story. Marfusha the rabbit, for example, launched in 1960 with Soviet space-dogs Otvazhnaya and Malek, and remembered in a series of haiku by punk rock singer John Talley-Jones of the punk band Urinals:

> Weightless rabbit flies
> through arcs of no gravity
> bathing in stardust.

> There's no hay in space
> so Marfusha must bring lunch.
> Chews her paper bag.

And a nod to the white cosmic radiation monitoring mousestronauts: A3326, A3400, A3305, A3356 and A3352, who accompanied Cernan, Evans, and Schmitt on the final Apollo 17 mission to the moon, and were re-christened Fe, Fi, Fo, Fum and Phooey. Right now on board the International Space Station, another group of mice (those hyper-intelligent pan-dimensional beings[*]) in the NASA Rodent Research Facility are conducting fiendishly subtle experiments on a rotating group of international astronauts, looking at the effects of long duration space flight on the human mind and body so we may continue our journey.

---

[*] See Douglas Adams on mice in The HHGTTG.

THE RIGHT SHEEP

Before we ever left the ground, animals tested the water first. On 19 September 1783, Joseph-Michel and Jacques-Étienne Mongolfier conscripted an unlikely group of subjects for a unique experiment. Their silk and paper balloon called the Aérostat Réveillon elegantly rose up from the Palace of Versailles in front of Louis XVI. The basket attached to it contained a cutting-edge experimental biological science package: a sheep called Montauciel (meaning 'climb to the sky'), there to approximate the physiological make-up of a human, as well as a duck and a rooster as scientific experimental controls. This historic flight lasted some eight minutes, achieving an altitude of 1500 feet and a distance of a couple of miles, before plopping down on the edge of a forest. The animals were unharmed, if a little flustered. The heroic sheep went into retirement living out its days in the personal zoo of Marie Antoinette.

To date there have been no sheep in space. History has seen other more practical species put forward as the advance guard in our exploration of the heavens.

The Ascent of the Aerial Balloon.

MOSS PIGLETS

A more subtle space pet is the tardigrade, also known as a moss piglet or water bear. At only 1mm, making them easy to carry, and considering that a 400g rat will cost you around $9,000 to get into space, they are certainly cheap to transport. Underneath the electron microscope they look suitably alien – like a tiny woodlouse in a spacesuit made out of a Hoover bag. They are the very embodiment of low maintenance being able to withstand temperatures as low as -272°C (-458°F, nearly absolute zero) up to about 150°C (300°F) and they are emotionally bulletproof. Tardigrades redefine our understanding of the parameters of life, which is why astrobiologists and children are so enamoured by them. They can go without water for ten years and food for thirty, and can survive doses of radiation that would certainly be lethal to most other life. They are the only animal known to be able to survive the harsh environment of the vacuum of space, no spacesuit necessary.* Entering a cryptobiotic state they can slow their metabolism down to 0.01 per cent, which is very handy for long trips. In 2007,

---

* And difficult to make presumably.

the European Space Agency's Foton-M3 (Tardigrades in Space or TARDIS) mission sent some desiccated tardigrades on a space mission, exposing them to the vacuum of space. When they returned to earth many survived and went on to reproduce, leading full, productive and happy lives.

BLOSSOMING – FRUIT FLIES

If you don't empty your compost bin or fruit bowl in your kitchen often enough, you'll probably see a few happy *Drosophila melanogaster* darting about. Fruit flies are the doyens of medical and genetic research: plentiful, low maintenance and crucially with a very short life cycle meaning many generations can be studied in a short period of time.

They also have the honour of being the first living creatures to leave the planet. After the Second World War, Nazi V-2 rockets captured by the Allies were tested at the White Sands proving ground in New Mexico. Part of this programme became Project Blossom – so-called because of the flower-like way the (often troublesome) parachute would open against the barren desert landscape. The nose-cone spaces of these rockets were no longer carrying munitions and so became used for carrying scientific equipment. On the 20 February 1947, rocket number 20 on launch pad number 33, with a scientific payload – containing some fruit flies to study the effects of cosmic rays in the upper atmosphere – was launched into the ionosphere reaching an altitude of 108 km. After the payload capsule was ejected, our heroic flies – spectacularly oblivious to their place in history – became the first life to cross the threshold into the void, before being captured once again by gravity's jealous hand. By all accounts, the flies returned safe and sound.

The next time you lift up a shrivelling half-lemon from your fruit bowl, consider for a moment that those fruit flies might be direct descendants of the first creatures in space. A back of the envelope calculation suggests your flies are 728 generations away. Give or take. In human terms that would take us back 20,000 years to the last ice age, so for the fruit fly their *one small step* happened in the late Pleistocene.[*]

SATELLITE DOGS

It's Kudryavka (Little Curly), sometimes known as Zhuchka (Little Bug), sometimes known as Limonchik (Little Lemon) who leads the long roll-call of Soviet space-dogs: a canine cosmonaut corps of strays plucked from the mean streets of Moscow. The chosen dogs had to be small – no taller than 35 cm and under 6 kg. They would be rigorously trained and lovingly cared for at the Moscow Institute for Aviation Medicine, principally under the watchful gaze of scientist Oleg Gazenko, a pioneer in aerospace medicine. Kudryavka's ascension to the heavens and sacrificial death is a familiar and potent narrative, and one that meant she became a true global icon, under the name by which she would be remembered forever: Laika (Barker), the first of the 'satellite dogs'. Unlike Lassie, that rough collie explorer of the parallel American moral universe, Laika represented a different set of values: national pride, sacrifice and Soviet strength.

Laika's suicide mission divided the world. The West, particularly in Britain, was outraged by the idea of deliberately killing a dog. The *Daily Mirror* shouted, 'THE DOG WILL DIE, WE CAN'T SAVE IT.' But in the Soviet Union, she was canonized as the first martyr of Tsiolkovsky's *cosmist* vision:

> **'Laika, sweet loyal Laika, how happy have you made the scientists all around the world! The faint beating of your heart, fluttering from a thousand kilometre altitude, to them has drowned out all other sounds.'**
> Olesya Turkina, *Soviet Space Dogs*

---

[*] Possibly.

80 ᠊

MONGOLIA · МОНГОЛ ШУУДАН

1978

The Sputnik 2 spacecraft in which Laika would fly was hurriedly assembled by Korolev's team to meet Premier Nikita Khrushchev's demand for a follow-up to the triumph of Sputnik 1. On a ferocious deadline, it was to coincide with the fortieth anniversary of the October Revolution which led, unfortunately, to flaws in the much more complex design process of building a vessel that could sustain the life of a dog. The *beep, beep, beep* of Sputnik 1 that shook the world would now be replaced by a living, beating heart. Laika was selected from a group of finalists, all of whom had been prepared for canine orbital missions. All were bitches – male dogs wouldn't have had the space to cock their legs. With electrodes surgically attached to her body and a waste collection bag harnessed to her, she was placed into the capsule (Object PS-2), a cushioned, walled box

big enough to lie, sit or stand in but not to turn around. Equipped with a food dispenser that would last for seven days or so, it was placed on top of the mighty R-7 Semyorka rocket in the space in the nose cone once designed to carry a nuclear warhead. Laika would have to sit patiently for three days on the Cosmodrome launch pad before lift off, as all the various systems were checked and issues ironed out. Finally on 3 November 1957 at 5.30 a.m. she left the earth, her heartbeat spiking from the violence and noise of the launch.

Within a few minutes a living creature was experiencing sustained orbital weightlessness, but it was sadly to be short-lived. The actual circumstances of Laika's

Above: **Laika**
Above right: **Oleg Gazenko holds Belka and Strelka**

death were only revealed years later, after decades of silence, misinformation and speculation. The truth was sadly much more gruesome than had been reported initially. She died a horrifically painful death within just a few hours, suffering from massive stress and overheating due to an inadequate on-board cooling system that failed to protect her against the heat from the engines and solar radiation that turned the spacecraft into an oven. By the third or fourth orbit the *beep, beep, beeping* of her heart had stopped. Laika in her metal sarcophagus would remain in a lifeless orbit for another 2,570 revolutions of the earth, a victim of political and engineering haste as much as of the vacuum of space, until on 14 April 1958 she was seen for the last time. A brilliant, bright, coloured object flashing across the sky with smaller bits breaking off was seen by observers on the east coast of America. What they were seeing was Laika's return home.

If Laika was to be cast as the martyr, Belka and Strelka, the space-dog double act, would soon grab the world's attention, becoming the first animals to be launched into orbit and return alive. The dogs who fell to earth became the first of the astronaut celebrities, but only by accident. They were replacements for Lisichka (Little Fox, Korolev's favourite dog) and Chaika (Seagull), who had tragically died in a launch explosion only weeks before. For this historic ride they would not travel alone. Keeping them company was a selection of mice, rats, rabbits and fruit flies packed on board the Korabl-Sputnik 2, the latest test vehicle for the new Vostok spacecraft. Belka and Strelka were launched on 19 August 1960. At first there was concern for their safety as the on-board television cameras showed no sign of movement, but after the first orbit they were seen to bark and they continued circling the earth for another seventeen orbits with Belka attempting to escape from her harness and suffering from space sickness. But sick or not, scientifically this trip was significant – it meant that weightlessness, dreamed of for so long, had in principle no serious effects, opening the door to Yuri Gagarin's Vostok flight less than a year away.

As well as the media frenzy and dog-based celebrity endorsements that were to follow, there was an interesting political twist when one of Strelka's puppies, Pushinka, from a litter she had after her space adventure was given to Jacqueline Kennedy as a gift from Nikita Khrushchev. The puppy was accepted by the Kennedys, having first been inspected and internally X-rayed for bugs or 'Doomsday devices'. Descendants of Pushinka are still with us today, while Belka and Strelka sit proudly, still and calm like two giant tardigrades in a cryptobiotic state, in the Moscow state museum.

Forty-eight dogs had been launched. Twenty had died. All were loved.

HAM – PRELUDE TO MAN

Monkeys, not dogs, was the more palatable choice of test animals to ride the rockets in 1950s America. As well as ammunition in the propaganda war – against the evil dog-killing Communists – there were other obvious reasons for using primates: intelligence, small size and similar physiology to our own. Project Albert began with a series of ill-fated early tests. A crude pressurized aluminium capsule was designed and constructed for a rhesus monkey that would fit into the nose cone of the V-2 rocket. On 11 June 1948, a 9-pound rhesus, Albert 1, was anaesthetized and sedated before being strapped onto a padded aluminium gurney that slid into the capsule. But even before launch there was no physiological data being transmitted or received – either there was a technical problem or Albert was already dead. Several other unfortunate 'Alberts' were to follow but none were to survive.

Throughout the 1950s, the experiments with monkeys continued, using the two-stage Aerobee sounding rockets.* As the race to be the first to get a human into space was heating up, a 7-pound rhesus monkey, Able ('A' for Able), and an 11-oz squirrel monkey from Peru, Miss Baker ('B' for Baker), were selected as two outstanding candidates for a technical space rehearsal. They launched from Cape Canaveral on an Army Jupiter missile on 28 May 1959 and reached an altitude of 360 miles (580 km). Nine minutes of weightlessness as they arced across the sky, before re-entry dealt them a bone-crushing descent. Joining them on the ride was a biological package that included fruit-fly larvae, human blood, mould, fish eggs, sea urchin shells and sperm. Both monkeys survived the ordeal, but Able died four days later after a post-flight medical procedure unrelated to the flight and is now stuffed and on display at the Smithsonian National Air and Space Museum in Washington DC. Miss Baker retired to the US Space & Rocket Center, gaining something of celebrity status before dying on 29 November 1984.

Ham and Enos were the two Astrochimp test subjects for NASA's dress rehearsal for the manned Project Mercury programme. Mission: to test the physiological, psychological and cognitive effects that a Mercury astronaut would be exposed to, as well as the performance of the spacecraft itself. Ham (male), 37 pounds, was trained and named after the Holloman Aerospace Medical Center in New Mexico where the candidates were intensively trained and subjected to radiation exposure, low pressure and cognitive tests. Launch day at Cape Canaveral was 31 January 1961. Ham had his pre-flight breakfast of cooking oil, flavoured gelatine, half an egg, baby cereal and condensed milk, and was then suited up and strapped into a form-fitting couch. But this was no free ride. On board he had to perform psychomotor tests, which he and all the other chimpanzees had been well trained for. In front of him in the capsule was a control panel with three lights and corresponding levers that were to be moved as prompted. Negative reinforcing with mild electric shocks to the soles of his feet would be administered if mistakes were made. The sub-orbital Mercury-Redstone rocket launched just before noon, and despite some teething issues Ham was unharmed, performing well with his 6.5 minutes of weightlessness and exposure to 17 g, demonstrating the stresses of spaceflight are not just tolerable, but that an astronaut could perform cognitive tasks under such stresses too. Post-flight celebrity and cultural iconic status didn't come easily to Ham and he showed no appetite to fly again, eventually retiring to the National Zoological Park in Washington DC and later to the North Carolina Zoological Park. He gained a lot of weight and at age twenty-six, weighing 175 pounds, he died of liver failure and heart enlargement.

---

* A research rocket: 'sounding' from the nautical term for throwing a weighted line overboard to measure sea depth.

LIFE

WHAT THE MONKEYS' RIDE TELLS US
AND PLANS FOR MAN IN SPACE

BIG RIDDLE FOR THE U.S. FAMILY:
WHERE DOES THE MONEY GO?

ICA'S SPACE
RAVELERS:
AND BAKER

JUNE 15, 1959

If Ham was a favourite of the handlers at Holloman, Enos had a less cuddly reputation earning the unfortunate nickname 'Enos the Penis' *not* because of his rumoured propensity to pleasure himself publicly, as has been enthusiastically circulated, but because of his general ill-tempered demeanour. Despite this, Enos was a remarkable chimpanzee and was selected to fly in preparation for John Glenn's orbital flight on 20 February 1962 despite having been pipped to the post by another primate, a Russian *Homo sapiens* called Yuri Gagarin, a few months earlier.

Enos had been trained for a gruelling 1,263 hours for the complicated and demanding mission, and was given a complex 'shape-based' psychomotor instrument panel, which would test his problem-solving abilities throughout the flight. The control panel was designed to spit out banana-flavoured pellets for positive reinforcement, as well as electric shocks on the soles of the feet when mistakes were made. The orbital rocket Mercury-Atlas 5 was launched on 29 November 1961 with Enos on board. The mission was planned for three orbits, but was cut short to two when among other problems Enos's control panel started malfunctioning by giving him seventy-six undeserved shocks. After 181 minutes of weightlessness he was brought back to earth, but had to wait for over three stressful hours at sea in the Mercury capsule waiting for rescue. Enos had clearly suffered, as the final NASA medical report shows:

In addition, the subject had broken through the protective belly panel and had removed or damaged most of the physiological sensors. He had also forcibly removed the urinary catheter while the balloon was still inflated.

Not only was this extremely painful, but it also caused a subsequent trauma and infection. Enos died almost a year later from a form of dysentery, a condition unrelated to the flight.

Ham is respectfully interred at the Space Hall of Fame in Alamogordo, New Mexico, although his skeleton, still government property, is unceremoniously laid out in a drawer in the National Museum of Health and Medicine in a suburb of Washington DC, labelled 'specimen 1871496'.

The post-mortem and whereabouts of Enos's remains are a mystery.

ASTROCHAT AND ASTRORAT

'Hardly had the shell been opened when the cat leaped out.'
Jules Verne, *From the Earth to the Moon*

Ardan, Barbicane, and Nicholl in Jules Verne's *From the Earth to the Moon* used a cat to first test the effects of the ballistic projectile's recoil. A squirrel had also been included, but had been eaten by the cat, becoming the first animal martyr to the space program.

In the early 1960s the French had been conducting their own animal rocket research in the remote Algerian Desert. Using the elegantly named Véronique* sounding rockets, a white rat called Hector was first to ride on 22 February 1961, a last-minute change to replace the original rat who had chewed through some cables thus forfeiting his spot. Hector, having been well trained on the rat centrifuge, and wearing his lace-up spacesuit which suspended him on a small metal frame, was inserted into the nose cone of the missile and thrown up to an altitude of 69 miles (111 km), returning to earth alive and none the worse for wear. Later Félicette, the Parisian stray cat, was selected from one of fourteen trainees, receiving a ten-hour operation to

---

* A combination of the name of the French town of Vernon, where a group of the Peenemünde German scientists went to work after the war, and the word electronique.

implant a skull electrode monitoring her brain activity that gave her a rather disturbing appearance. She was launched on 18 October 1963, went 97 miles up for about quarter of an hour, and was recovered successfully. To date Félicette is the only confirmed cat to earn her chastronaut wings, for good reason: as we know, cats don't respond well to being herded.

Top: **From Jules Verne's *From the Earth to the Moon***
Above: **Hector the rat and a trainee space cat**

IVAN IVANOVICH – THE RUSSIAN DOLL

A month before Yuri Gagarin's flight on 12 April 1961, in a test of the Vostok capsule (Korabl-Sputnik 4, known in the West as Sputnik 9), sat the quiet, obedient and little-known steely eyed missile man Ivan Ivanovich (the Russian equivalent of the American name 'John Doe'). Before there was a man, there was a mannequin. The 'phantom cosmonaut' Ivan had flexible joints, leathery skin and a detachable head, and was dressed in the same SK-1 orange pressure suit that Gagarin would wear – with the words 'MAKET' (Dummy) covering his face to avoid any identification mix-up on discovery by an unsuspecting Russian farmer. His human-shaped body cavity, just like a Russian doll, became a spacecraft within a spacecraft housing a bizarre assortment of living things – an anthropomorphic ark that not even Tsiolkovsky could have conceived of in his wildest dreams. Inside were crammed some eighty mice, guinea pigs, reptiles, human blood samples and cancer cells, as well as various technical instruments, including a tape recording to test the communications featuring a Russian choir and a person reading aloud a recipe for borscht. By his side in a separate pressurized capsule was Ivan's faithful space dog Chernushka ('Blackie'), a mannequin's best friend. For his second mission, Korabl-Sputnik 5, he was accompanied by Zvezdocha ('Little Star').

Before humans had left the planet, Ivan Ivanovich was there first. A shadow cosmonaut. Without life, yet full of the lives of perfect creatures.

Opposite: **Close inspection of Ivan Ivanovich in a New York auction warehouse.**

ZOND 5 – THE TORTOISES WHO THOUGHT THE WORLD WAS FLAT AND THEN SAW IT WAS ROUND

On the evenings of 19 and 20 September 1968, just three months before William Anders, James Lovell and Frank Borman set out on their historic Apollo 8 round trip to the moon, the Jodrell Bank radio telescope in Cheshire intercepted a mysterious human voice coming from deep space. Even more bizarrely, the crackly broken voice seemed to belong to Russian cosmonaut Valery Bykovsky and he seemed to be reading out telemetry data. The giant radio dish was tracking the Soviet spacecraft Zond 5 (Zond is Russian for 'probe'), which was returning from its close-up look at the moon. Had the Russians got a man to the moon first while the Americans were napping? Not quite. At first the Soviet news agency TASS denied the existence of the spacecraft, before confirming it a few days later.

Unlike earlier Zond probes, which were designed for planetary exploration, the later (and misleadingly named) Zond spacecraft were variants of the Soyuz, minus the orbital module. Zond 5 had been launched from the Cosmodrome aboard a Proton rocket on 15 September. Its mission would be a historic first: testing the water for a manned spaceflight to perform a lunar flyby and return safely to earth. What made Zond 5 particularly interesting was the ragtag crew of travellers on board that were about to make history. Strapped into the pilot's seat was another 70kg human mannequin with radiation measuring devices, a tape recorder with cosmonauts' voices to test the radio communications in deep space and cameras to record the journey. But more important were the mannequin's companions – as well as the mealworms, fruit-fly eggs, plants, seeds and bacteria, in command of this unlikely ark were a pair of Russian Steppe tortoises.

If ever there was an animal designed by the prudent and economical hand of natural selection to be an astronaut, then it's the hardy Russian Steppe tortoise. Or to give

them their space name, *Testudo horsfieldii.*
Tough, methodical, stoic, low maintenance,
easy to pack. With thick skins and full
protective body armour, they're nature's
perfect space pets. Not for nothing was the
humble tortoise elevated to star billing in
Aesop's famous fable. In the dying months
of the race to the moon, with the Americans
sprinting to the line, what animal could have
been more appropriate and symbolic? The
tortoises were packed on board twelve days
before the launch, having had their final
meal on earth. After a successful launch the
mission suffered an early setback, as the star
tracking navigation system failed to work,
but limping on the strange craft came to
within 1950 km of the surface of the moon
on 18 September before swinging around
and heading home. The small porthole was
filled with the view of the earth, witnessed
by the on-board cameras, but meaningless to
the creatures on board. The humble tortoise,
whose entire understanding of the cosmos
up to that point was the perfectly flat horizon
of the Kazakhstan steppe, was witness to the
earth as a spherical blue marble, suspended
in a starfield.

Our first living trans-lunar explorers
survived the terrifying sustained 16-g re-entry
ordeal, having lost some 10 per cent of their
body weight, but arrived home with apparently
good appetites. The race to the moon had
been won by the tortoise. History does not
record their names. The nimble hare (Apollo 8)
crossed the finish line three months later.

Opposite: **Earth photographed by Zond 5**

HOW TO GO INTO SPACE WITHOUT LEAVING HOME

analogue: a person or thing seen as comparable to another; 'an interior analogue of the exterior world'.

MOON GOOSE ANALOGUE:
LUNAR MIGRATION BIRD FACILITY

'What happened to the moon geese in the twenty first century?'
Agnes Meyer-Brandis

Below the mountains in the Abruzzo region of central Italy, amid the woodland, olive groves, junipers, and grazing sheep, sits a large rural farmhouse-cum-international-research-station called Pollinaria. Pollinaria explores the fertile intersections of art, poetry, science and traditional methods of agriculture. In 2011, the place became base camp to a very special astronaut training project. German artist Agnes Meyer-Brandis breathed life into Francis Godwin's 1638 lunar adventure, by incubating and hand-rearing eleven of the forgotten species of migrating moon geese. Each of the geese was named after a space pioneer, the names written on their eggshells in pencil. On hatching, Agnes 'imprinted' the newly born astro-chicks' behaviour using strong visual cues: she would wear a home-made silver spacesuit and yellow wellington boots, which the geese would follow as if she were their mother.

For nine months the flock followed Agnes everywhere she went as she conducted her own unique moon-goose astronaut training programme. The migration training period was followed by habituation to a lunar-surface analogue in a purpose-built facility (outhouse) preparing them for life on the surface of the moon. Perhaps next time you're on holiday you could devise your own space mission. Or just lie back at night, and let your mind migrate to the stars. No one will ever leave the planet quite as beautifully as this.

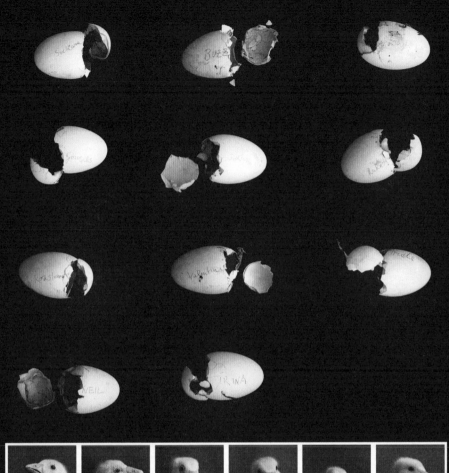

Neil

Svetlana

Gonzales

Valentina

Friede

Juri

Buzz

Kaguya-Anousheh

Irina

Rakesh

Konstantin-Hermann

Moon Goose Colony
age: 11 days
born May 11-12th 2011

shaping processes as Mars and our planetary neighbours, and of course we share the same chemistry and laws of physics. Mars is a stone's throw away from you: the sedimentary formations of the North Berwick coast, or the volcanic island of Lanzarote where ESA runs the PANGAEA programme training astronauts in the basics of planetary geology; the red deserts of central Australia, or the hyper-arid Atacama Desert in Chile. Other places are the Grand Canyon, or the famous Wadi Rum in Jordan where Ridley Scott filmed *The Martian*. If you can willingly suspend your disbelief, with the right photographic filter you're as good as there. The famous Martian photograph of Carl Sagan standing next to a model of the Viking Lander was shot in Death Valley. As a child this image always confused me.*

LET'S GO TO UTAH

If you want a deeper, more immersive experience then why not join an analogue mission. The next time find yourself in the American south-west, drive north-east across the rugged Martian-looking landscape from Las Vegas. Type 'Cow Dung Road, Utah' into the satnav and you will be taken far off the beaten track, to the Mars Desert Research Station. It's a large white round metal spaceship-like building called The Hab, with orange landing legs sticking out of the side, which looks as if it has landed on the red planet. Next to it is a long cylindrical-shaped greenhouse facility, and a few yards away on a rocky escarpment is the Musk Mars Desert Observatory telescope. The rugged multi-hued landscape of orange and red is indistinguishable from the extraordinary photographs collected from the NASA Curiosity Rover currently inching its way around Mars' Gale Crater, its wheels now cracked and worn from five years of Martian exploration.

SPACE STAYCATION

Leaving the planet is a massive and expensive faff. An alternative is to go into space or visit another world, without the bother and expense of actually going into space or visiting another world. Space analogues recreate on earth, with varying degrees of fidelity, the physical and psychological conditions of a space mission, or of an extraterrestrial environment, like the surface of the moon or Mars.

If it's just a quick snapshot view of the Martian landscape you're after, many parts of the world have a physical geography or geology that closely resembles Mars. The flotilla of spacecraft sent to Mars have revealed a landscape as familiar as our own. We share many of the same geophysical-

---

* Sagan had borrowed the Viking replica for his TV series *Cosmos* on the promise that he wouldn't damage it. Filming in Death Valley seemed sensible, given that it's one of the driest regions of the world. Ironically they were met by a heavy rainstorm – hence the orange anorak, which he hastily grabbed from the front seat of a crew car and then had to keep wearing for continuity. It went on to become one of the defining images of the series.

This is one of several Mars analogue facilities operated by the Mars Society, including the Flashline Mars Arctic Research Station (FMARS) on Devon Island in the Canadian high Arctic. A two-week stint living as a Martian will set you back $1000 or so. Design yourself a mission patch, give yourself an excellent mission name and a website, and do some proper research – from your base you can explore the surroundings in your spacesuit and bubble helmet on foot or by driving around on your quad bike. You can collect pristine soil samples, look for signs of life, start a new religion, conduct fiendishly revealing psychological experiments on your crew mates, or try to grow potatoes using your own faeces as fertilizer. Until someone comes from the outside world to rescue you.

Opposite: **Carl Sagan filming *Cosmos***
Above: **Dr Melissa Battler, Mars Desert Research Station, Utah**

LONG DURATION MISSIONS

The Mauna Loa volcano in Hawaii is home to the HI-SEAS (Hawaii Space Exploration Analog and Simulation). Crews of six are chosen with the same strict criteria that might be used to select astronauts – high-functioning engineers, astrobiologists, medical doctors, computer scientists and pilots are the order of the day. A large solar-powered white geodesic dome sits like a ripe pimple at over 8000 feet on earth's own mini version of Olympus Mons* and will be your home, isolated from the rest of the world, for up to a year. Your only contact with the outside world is online, perhaps with a 'Martian delay' to simulate the real earth–Mars time delay thrown in for good measure. Through the dome's airlock you'll emerge wearing your spacesuit, to explore the ancient volcanic Martian landscape, dreaming of the earth you've left behind.

Here, the primary aim is to explore and experience the interactive behaviour that crews will experience during long duration space missions. How can a group of people live together in a confined space isolated for months without going mad, becoming depressed or killing each other? What makes a good crew? What is the effect of food and its preparation on morale? How will sleep patterns develop without the natural earth day/night cycles? What would happen if you really hated someone? Or fell in love? Or became jealous of someone else in love?

Not all Mars analogue studies offer the chance of escape to wander the great outdoors. The daddy of them all was the Mars500 mission, which took place in a specially built facility of connecting sections resembling the International Space Station, located in a large hangar at the Institute of Biomedical Problems in Moscow. For 520 days (the length of time for a Mars mission) the crew consisting of three Russians, two from ESA and one Chinese – were locked away from the outside world. The mission, among many other things, highlighted the reality of international crews living and working together. Different cultural holidays were observed and shared between the group, like Chinese New Year, Christmas and Halloween. For seventeen months they ran experiments on themselves and each other. Artificially delayed communications were built in to replicate the time delay between Mars and earth. Using adapted spacesuits, emergency Mars surface EVAs were conducted in a specially controlled indoor Mars yard, along with studies on sleep patterns and depression. As much as anything it was a study on the effects of monotony, something

---

* Biggest volcano in the solar system on Mars.

you're going to have to come to terms with if you're planning on going anywhere past the immediate vicinity.

News from the outside world was scarce, and what made it through became cherished moments to cling on to. One piece of news they heard in 2010 concerned the fate of the 'Los 33' Chilean miners, who became trapped 2300 feet underground after a cave-in at a mine near Copiapó in the Atacama Desert. In solidarity with their predicament, the Mars500 crew decided to sit down and write them a letter explaining the story of their own self-imposed isolation, why they were doing it and wishing them well. ESA took it upon themselves to get the letter to Chile and managed to hand-deliver it to the trapped miners via the ventilation shaft.

Opposite: **The Mars500 facility in Moscow**
Above: **Jean-François Clervoy's underwater moonwalk**

UNDERWATER ASTRONAUTS

Of course, all these studies cannot recreate the absence or change of gravity and the effects of radiation that a genuine space mission will entail. Analogue facilities like NEEMO (NASA Extreme Environment Mission Operations), off the Florida coast near Key Largo, in which astronauts live and train underwater, and the long-duration bed-rest studies like the Envihab laboratory in Cologne, try as closely as possible to replicate the gravitational effects of muscle and bone loss that have become the defining physiological effects of long-duration space flight.

One particularly novel underwater space analogue was ESA's recreation of the Apollo 11 moonwalk: Apollo 11 Under the Sea, with underwater astronaut Jean-François Clervoy (Neil Armstrong) and astronaut trainer Hervé Stevenin (Buzz Aldrin) wandering around the bottom of the Mediterranean. Their suit buoyancy was tweaked to give them the effect of one-sixth of Earth's gravity. They collected soil and rock samples using similar lunar geological equipment, and more importantly planted a European Flag. You and a friend could learn the lines from the historic Apollo 11 EVA, and carry out such an experiment yourselves.

All of these examples can only approximate to various degrees the demands on mind and body of space travel. And one thing they share is the fact that the door through the wardrobe back to the real world can always be opened. The air outside in the real world will be breathable. You can always swim back to the surface. The inherent catastrophic danger of being millions of miles from earth is removed. Even the International Space Station has a lifeboat* attached to it if things get really bad.

WHITE MARS

'Polar exploration is at once the cleanest and most isolated way of having a bad time which has been devised... Take it all in all, I do not believe anybody on earth has a worse time than an Emperor penguin.'
Apsley Cherry-Garrard,
The Worst Journey in the World

Analogous comparisons between Antarctica and space exploration are many and useful. Humans have poked and prodded both realms and, while not yet fully adapted to either, we now have a continuous presence on the fringes of both – international settlements on the Antarctic coast and our international outpost in low earth orbit. The continent of ice and the realm of space are joined in spirit and in law by two legal treaties, the Outer Space Treaty and the Antarctic Treaty, which prevent colonization and exploitation.

The International Geophysical Year of 1957–8 didn't just see the Soviet conquest of space with the launch of Sputnik 1, it also saw

Soviet expeditions push into the high region of the Antarctic Plateau and the founding of the Vostok research station, which holds the record for the lowest reliably recorded temperature on earth at -89.2°C. The scientific research train of the third Soviet expedition, having travelled 2110 km from the coast, stopped at 82°06' S 54°58' E – the most remote point on earth, and the furthest point in Antarctica from the sea. There they set up a small research station. This is the Southern Pole of Inaccessibility. If you truly want to get away from it all, this place should be first on the list. In summer, these places are relatively accessible (cosmically speaking) by air and traverse, but the nine months of winter means permanent darkness, with temperatures similar to those on the surface of Mars, nudging -100°C. No aircraft could rescue you in these fuel-freezing temperatures. The winter conditions make being here the closest you might feel to standing on another world without getting off this one.

That said, if you had to pick the very best spot on earth for your space adventure, how about a trip to a place known only as Ridge A (81°30' S, 73°30' E)? Ridge A is where the 'stratosphere comes to the ground' and is about 90 miles south-west of Dome A (Dome Argus) on the Antarctic Plateau. Ridge A has been identified as the best place on earth for astronomical observations and is home to the remotely operated HEAT (High Elevation Antarctic Terahertz) far-infrared telescope. It is ultra-bone-dry, lifeless, completely calm and free from any sort of weather, sitting in the eye of the storm at the still centre of the polar winds that swirl around the plateau. Come here in the middle of winter, when the sun never rises, with twenty-four hours of blackness and no sound other than your quickening heart. Like a glass-bottomed boat, you can look out into the universe. Here, as in space, the stars won't twinkle. Take it all in. It may be the last thing you ever do.

* Soyuz.

TRAVEL GUIDE:
THE HIVERNAUT

Name:
Beth Healey
Profession:
European Space Agency Medical Doctor
Claim To Fame:
Winters over in Antarctica

You've overwintered in Antarctica? That's a big deal. I was working for the European Space Agency at Concordia, which is the space analogue programme at Antarctica. Concordia is specifically looking at long duration space flight. We're really interested in the isolation you get there. For nine months of the year you're completely alone, and that puts certain psychological and physiological pressures on the crew. That's a lot like what we're going to have to do for long duration space missions of the future.

Is Antarctica the closest thing to going to space that we can get on earth? I believe so, yes. It's the only place where we can truly and ethically isolate people. There are lots of analogue programmes like HI-SEAS, and Mars500. But the thing about a programme like that is essentially there is a door that you can walk through at any point and, after having been to Antarctica, I think I would have found that a lot harder. I would probably have spent 499 days of the 500 days wondering whether or not I should leave the programme. In a place like Concordia you really are truly isolated, so that's not really a concern. Whether it's harder

or easier I'm not really sure, but it certainly put a different psychological pressure on you because you're not really thinking about leaving.

Lots of people go to Antarctica to do science, but generally that's in the summer.
Exactly. There's a huge population of scientists going to Antarctica during the summer months that runs between November and February typically. And a lot of the stations are coastal stations. That's the big difference – Concordia is one of only three inland stations, along with the Amundsen-Scott South Pole and Vostok stations.

Being there in winter makes you a hivernaut? How did this happen to begin with?
Officially I'm a hivernaut, yes! From a really early age I was going off and doing adventurous things with my dad. When I was three or four I had my own little kayak. It wasn't an unusual thing for me to get involved in expeditions when I was growing up. I became very interested in remote medicine and working as part of medical logistical support teams in remote environments. I had already been to Greenland, Siberia and the North Pole, and so going to Antarctica was something that I was really interested in doing.

What happens if somebody gets ill? On the Space Station you have a Soyuz lifeboat attached to it so you could come back down to earth if the worst happens.
That's why Concordia's used. We had two medical doctors: the base medical doctor and myself. I was the ESA doctor. Between us we could do most surgeries. We also had a fairly sophisticated telemedicine, which is something we are looking at for long duration space flight. For the whole time we are down there, we have direct links to a hospital in Rome where we've got links to all the different specialists, so we are able to perform most different types of surgeries.

Let's say if someone gets an aggressive form of cancer for example?
You wouldn't get them out. There's no chance. That happened at the South Pole and they air-dropped stuff in. With Concordia, it's higher, it's even colder and they have never done it.

Tell me about the psychological test that one has to take in order be a hivernaut.
It was all a huge blur… it was a three-hour interview talking about every aspect of my life, my family situation. We had to do the classic Rorschach test where you look at the cards and tell what you see. I tried to see happy thoughts!

Death? Murder?
I did say at one point I saw a dead animal, and I was like 'Oh god! I wonder if that's jeopardized my chances.' It did look like a roadkill zebra so I think that was fair enough.

Is hivernaut training like training as an astronaut?
Yeah, there's lots of similarities. Because you are in many ways like an astronaut – you're doing in-the-field data collection for lots of other institutes. That's exactly what they're doing on the ISS – it's the same kind of model. In addition to that we had other training. Because I'm a medical doctor I went out to Chamonix.

Skiing? Very important.
… skiing, and more importantly I did mountain rescue medical training. As the ESA doctor your primary role within the system is to go out and rescue the person, bring them in to the base doctor who's going to be setting up the hospital, in terms of doing operations and such.
   And then we also did human behavioural performance training as crew.

HIVERNAUTS AND LOVE
You're in Concordia, and it's been summer and it's all nice and it's not too difficult. And then…
There's loads of stories about what it's going to be like, before you go, so it's quite

a weird time. That there's just going to be people going crazy, committing suicide. No one's actually committed suicide but people have attempted it. Nobody was like that in our crew, but it's the isolation. And in that particular case I think it was a relationship issue between one of crew members and them. She 'got' with somebody else, basically.

Being a human is difficult, especially with things like relationships.
And being a girl, a lot of people thought I was an experiment, to see what would happen, by putting a young girl into the crew. Which is weird.

Did they tell you that?! They actually thought you were an experiment?
Yes, this one person in particular. He'd already done three overwinters and he actually wrote to ESA and the Italian Antarctic programme saying he thought I should be sent home because I'm too pretty.

You're joking? Wow, that's a backhanded compliment.
Yeah, I think he thought he was doing me a favour in terms of – it would destroy me, going to Antarctica and the overwintering experience, because I think he thought it was too hard and it would get too much… Not so much me, but he thought I would

disrupt the crew. That it would cause problems for other people.

And were you in a relationship with anyone at this point?
No.

And was he one of the people who ended up wintering with you?
No, he'd just finished wintering. This was the same person who also had slugs as pets.

What? Slugs as pets?
The Antarctic treaty means that you're not allowed to bring in any foreign animals. He'd been there fourteen months, and with all the fresh fruit and vegetables coming in you get some slugs, and he actually kept them as a pet as company. I certainly could appreciate why he did that after I'd overwintered, but when I'd just arrived I thought, this is just a crazy place.

But there was also a lot of weird stuff that happened when I first arrived, like people put my Ugg boots in the fridge and they tried to shrink all my washing.

How do you know that wasn't you being paranoid? Why would someone put your Ugg boots in the fridge? At no point in my life would I consider putting someone's Ugg boots in the freezer.

I know, it's weird, isn't it? I don't think they did it by mistake.

But things become very precious in these environments. It's a lot like space travel. Somewhere like Concordia, it's the one true place where money doesn't matter, everything is free, alcohol is free, you're all given exactly the same clothes and it's a true playing field. It's a bit like prison in that respect and there's no sort of social hierarchy. If anything, I would say scientists are almost looked down on because you're not really essential in terms of survival within the crew. So somebody that perhaps organizes water, or does the power station, you know, the real kind of survival jobs in the hierarchy of the station are much more valued.

A lot of people were hostile about the idea of coming back out of the isolation phase because they were happy with how things were, and they didn't want to go back home. We had people like submariners, for example, and they found the overwintering experience a lot easier than being in the real world.

In some ways it is easier – you've got a chef there, you've got your science to do, but you haven't got the rest of the faff of real life to think about. You don't have to cook for yourself, you don't have to buy anything, you don't

have to think about what you are going to wear. You just have to do your job; it's a very simple life.

It sounds like there were quite a lot of interpersonal problems.
It's a tough environment. I think one of the big reasons is the language barrier, everyone speaks different languages and when you're really knackered you kind of just want to speak your own. So I think that was one challenge. I think being a girl has different challenges.

How many women were there?
Three.

How do the agencies study or measure the psychology aspects?
We were wearing activity watches and they were obviously looking at activity levels and sleep/wake cycles but also crew interaction. My watch would interact with your watch and it would also know where I am on the base. With the activity watches it's looking at how relationships are changing over time and also personal preferences, so – are you seeking social interaction, are you in a social zone on the base, or are you isolating yourself in your bedroom, are you choosing to not interact with other people, and also looking at how the group dynamic changes over time.

If you fancy someone?
People fall in love, people fall out of love, get jealous. Relationships must happen.
It's a funny thing. People would be actively... unfriendly to me, because otherwise they would get teased that they fancied me. It's a minefield. If you were friends with someone, everyone would assume you're sleeping together.

Presumably people were sleeping together. If there were thirteen people locked up for nine months.
Not so much. In the summer a lot more, but during the winter... I think because you are there as a crew, unless you really, really like somebody and you think it's going to work for the whole winter I think it's too high a risk.

SPACE WALKING
Did you have a worst day?
I had one or two. I remember one moment where it was the worst day. I think it was just a day where I was really, really tired. Obviously, with the winter you don't sleep well. For 105 days we don't have any sun. Permanently dark.

Do you go outside? Is it too cold? When you say dark, is it pitch-black? Can you see the stars the whole time?
For a month you could not tell if it was day or night. You see the stars clearly, the

Milky Way is there all the time. You see the Aurora all the way through the day.

How far are you away from another human being?
Six hundred kilometres is Vostok Station.

Six hundred kilometres – so actually the six people on the space station are probably as close. If you went outside what would happen? How long can you go out?
Concordia is typically -70°C to -80°C. It's really, really cold. You have to wear all the gear. They say it's a lot like a spacewalk because you have to be totally covered up. You have to really think about going, you're taking radios out and tell people where you're going and you normally go with another person. It's dangerous. You have limited dexterity; if you get cold it's hard to open up doors. We have shelters outside which are heated, so you're never far away from a safe place I suppose, but it is quite extreme.

How did the physical landscape affect you?
It's a lot of sensory deprivation as well, you don't have any smells, that's one big thing. It's like space in that respect. You don't have any mud or dirt. You don't have any human smells. Nothing smells of anything.

Did it change you and the way you think about the world and the way you think about life? You come with expectations and the expectations become reality. I think I did change. Living without the sun really disconnects you from the real world. Having the sun in the sky always connects you to wherever you are. The one thing I really noticed when you didn't have that sun is that you really did feel like you were on a different planet. It was strange and exhausting, because you didn't have that stimulation of the sun to wake you up ever.

Did you ever go outside and just look up? Yes. And you become deeply connected to the universe. The stars are everywhere. It's beautiful. You're much more aware of everything. You just feel much more connected with the outside, and with what's going on, and the temperature. Things that I would never think about when I'm back here in civilization.

What the trapped Chilean miners made of the letter writing gesture has never been reported and a reply was never expected by the Mars500 crew. But the act of writing a letter of solidarity, or the act of putting someone's Ugg boots in the freezer, are motivated by a recognition of something of deep importance. Humans are a complex social species. The fear of isolation, of separation and confinement in all its many forms both physical and psychological, is deep rooted and the effects of those fears manifests itself in many ways. It is one we will have to work through if we are to sever our earthly connections. What kind of people can deal with these stresses? As earth recedes out of sight, as we sail further into an empty universe, what will we be left with? A memory of the smell of the grass of home and the terrifying prospect of only ourselves for company.

SECOND STAGE:
CLEARING THE TOWER

'Between you and infinity there are billions of stars'
Leonard de Vries, *The Second Book of Experiments*

DO I HAVE THE RIGHT STUFF

“

AL WORDEN, APOLLO 15 COMMAND MODULE PILOT, ON THE MOMENT OF LIFT-OFF

Talk us through that moment when you’re sitting on top of the Saturn V, the biggest rocket that’s ever been built, and you’re sitting right on the top about to go.
We got in our spacecraft about 7.30 in the morning for a 9.30 launch. We got out there and of course it was dark. We got inside the spacecraft and it was all chilled down. The air-conditioning system was down low, -45°F, as a matter of fact, that’s what it was set at. We couldn’t see outside, it was dark inside. All we had was an instrument panel to see, and we were just sitting there and I went to sleep. What else are you going to do? You’re waiting for the next thing to happen. It’s pretty doggone boring, just sitting there doing nothing, I’ll tell you.

Domingo Gonsales didn’t go to sleep, he was having a fantastic time...
He didn’t have a chance to sleep. He had to control those geese...

”

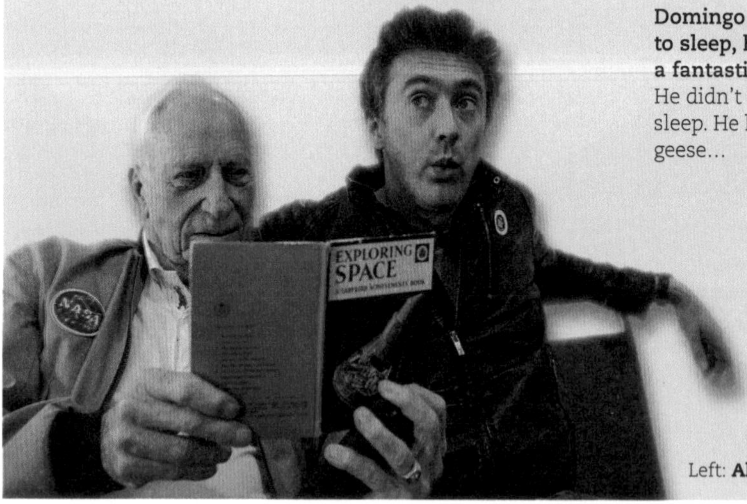

Left: **Al Worden and author**

The famous Austrian psychiatrist Viktor Frankl tells us:

Between the stimulus and the response there is a space. And in that space lies our power and our freedom.*

Whatever the 'right stuff' is, it lies somewhere in that space.

The reality of space travel has meant that astronauts have some pretty clearly defined qualities and characteristics. At some point you've probably wondered about whether you have this 'right stuff' – 'that indefinable, unutterable integral stuff'. In theory, anyone, including you, could climb on board a rocket and take off, and unless it blew up you'd still

be alive. But it's how you behave and cope with whatever the universe throws at you when you're up there that's the question. When the hull of the space station is breached and you hear the slow hiss of air venting into space; when your spacesuit helmet fills with water during an EVA; when the Saturn V rocket experiences an unknown electrical failure on launch. Your hand is on the abort handle. Do you turn it? You have a gap between stimulus and response in which you are in control. How will you respond? With what kind of *stuff*? Praying will only get you so far.

'Dear God. Please don't let me fuck up.'
Alan Shepard's test pilot prayer**,
Tom Wolfe, *The Right Stuff*

ARE YOU CHUCK YEAGER, TEST PILOT & 'RIGHT STUFF' LEGEND?

Tomorrow morning you're about to fly faster than any human has flown before in an experimental rocket plane (Bell X-1), to explore the very limits of what is considered even possible. What you choose reveals a lot about you...

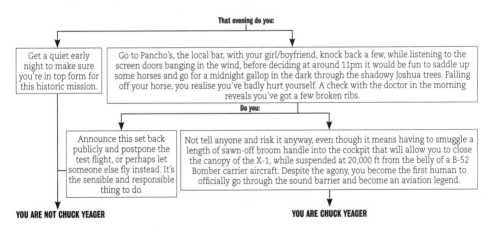

That evening do you:

Get a quiet early night to make sure you're in top form for this historic mission.

Go to Pancho's, the local bar, with your girl/boyfriend, knock back a few, while listening to the screen doors banging in the wind, before deciding at around 11pm it would be fun to saddle up some horses and go for a midnight gallop in the dark through the shadowy Joshua trees. Falling off your horse, you realise you've badly hurt yourself. A check with the doctor in the morning reveals you've got a few broken ribs.

Do you:

Announce this set back publicly and postpone the test flight, or perhaps let someone else fly instead. It's the sensible and responsible thing to do.

Not tell anyone and risk it anyway, even though it means having to smuggle a length of sawn-off broom handle into the cockpit that will allow you to close the canopy of the X-1, while suspended at 20,000 ft from the belly of a B-52 Bomber carrier aircraft. Despite the agony, you become the first human to officially go through the sound barrier and become an aviation legend.

YOU ARE NOT CHUCK YEAGER

YOU ARE CHUCK YEAGER

---

* Like many great quotations, he didn't actually say this. See end notes.

** He didn't actually say it quite like this.

THE RIGHT ORIGAMI

The Japanese Space Agency JAXA have perhaps the most brutal right stuff test for their astronaut candidates. It's a searching examination of character. Candidates are observed in isolation for a week and must quietly and methodically, in the time allotted, fold a thousand origami paper cranes* while trying to avoid any deterioration in their craftsmanship. The paper cranes are taken away and analysed to see if they have varied over time, which would be a sign of impatience under stress. The one thousandth crane should be as precise and elegant as the first.

THE RIGHT NATIONALITY

Most of the qualities that will affect your route to becoming an astronaut will depend on things beyond your control, like your nationality for example – does your country have an active human space programme? If you're from America, Canada, Japan, Russia, Europe or China then you've already got an advantage. Of course, there are ways around that. The NASA astronauts Mike Foale and Piers Sellers were both born in the UK, which at the time had no investment in human space flight, but had American citizenship. Remember, you could always marry someone from the nationality of your choice. Preferably the person who is in charge of choosing astronauts.

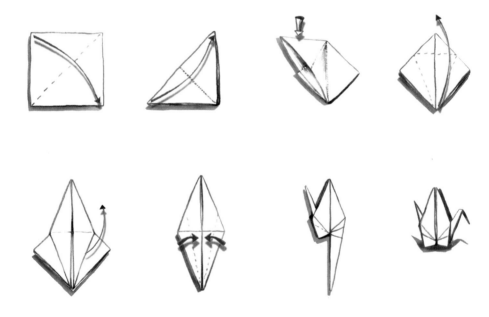

* A thousand cranes is a Japanese symbol of good luck.

Where you're from is going to determine what you're called too. 'Astronaut' from the Greek 'Star Sailor', if you're American or European, although the French space traveller is called a 'Spationaut'. 'Cosmonaut' if you're Russian, 'kosmos' meaning 'universe', and 'Taikonaut' if you're Chinese, 'Taikong' meaning 'great emptiness'. And who could forget the 'Afronauts' – Zambia's 1960s space programme, with their rudimentary training methods and rockets made from oil drums, run by the antic visionary Edward Makuka Nkoloso.[*]

Your next problem is: who's hiring? At the time of writing, the latest NASA group of astronaut candidates has just been announced, while ESA's last group, known as the 'Shenanigans', was picked in 2009.[**] Canada have been hiring this year. You just have to keep checking your local national space agency websites.

THE RIGHT LIFE SKILLS

Then there are the things you can do something about: are you a good swimmer? Do you have 20:20 vision? These were both problems that faced NASA astronaut Mike Massimino, but he made sure he got them sorted. Can you fly an aircraft? Are you in the military? Do you have a doctorate, or a degree in a STEM (science, technology, engineering and mathematics) subject, or just a 2:2 in English and Drama? Money can help bypass many obstacles. Are you an Internet billionaire, for example? Being an Internet billionaire generally helps. The entrepreneur Richard Garriott and others managed to buy a trip to the ISS, but it's not cheap. What about your personality? I've never met an astronaut I haven't liked. There's a special thread that runs through all of them, a temperament that they have in common. Like the stars seen from earth they have a twinkle, but like stars seen from space they are constant and unwavering.

Being chosen as an astronaut is only the beginning of the story. What follows is years of training, the constant threat of being grounded for a million reasons and all this with no guarantee of a flight.

THE RIGHT STUFF FOR WHAT?

The job description of the astronaut is constantly changing. The Project Mercury capsule that graces the front of my 1960s Ladybird *Exploring Space* book was designed for a single test pilot, like a human cannonball, to briefly poke our heads above the atmosphere, to show we could do it. The Gemini spacecraft was designed for two men working together for several days, and involved egress from the spacecraft and docking manoeuvres – a rehearsal for Apollo. The Apollo astronauts had to be test pilots, explorers and scientists, visiting a new world. The Shuttle astronauts had to be electricians and telescope repairmen too – fixing the broken Hubble Space Telescope became one of the Shuttle era's defining missions. And now on board the International Space Station, days are spent running a battery of scientific experiments, as well as conducting public relations exercises and carrying out the symbolic role of maintaining our remotest human outpost. Perhaps with the advent of space tourism the astronaut's job will be simply to have a good time and enjoy the view.

But for now at least, the selection and training remain rigorous.

---

[*] See page 100

[**] Matthias Maurer was a finalist in the 2009 selection, but was only made an astronaut in 2015.

THE RIGHT START

President Eisenhower created NASA, the National Aeronautics and Space Administration, specifically as a civilian organization whose purpose was to wrestle the control of space exploration away from the military. NASA opened its doors for business on the 1 October 1958. A few weeks later, a band of engineers led by Robert R. Gilruth called the Space Task Group was created to formally instigate a manned space programme. With it, Project Mercury was born, named after the Roman god of travellers, eloquence, trading, thieving, skill and luck. Top of the agenda was the selection of the newly termed 'astronauts'. A group of flight surgeons, psychologists and psychiatrists would identify and hunt down America's finest humans. It's here we see the astronaut's job description clearly defined for the first time:

1 To survive; that is, to demonstrate the ability of man to fly in space and to return safely

2 To perform; that is, to demonstrate man's capacity to act usefully under conditions of space flight

3 To serve as a backup for the automatic controls and instrumentation; that is, to add reliability to the system

4 To serve as a scientific observer; that is, to go beyond what instruments and satellites can observe and report

5 To serve as an engineering observer and, acting as a true test pilot, to improve the flight system and its components

THE RIGHT SEARCH

Pilots, balloonists, submariners, deep-sea divers, scientists, polar explorers and mountain climbers were all thrown into the mix. YOU might have been just what they were looking for. It was Eisenhower himself who made the executive decision that the first group of astronauts would be selected from America's finest test pilots. As well as the obvious skills and qualities that test pilots possessed

in flying experimental high-performance aircraft, there was also the ease of access to military records and national security issues. It was decided that the pilots would need to have a minimum of 1500 hours flight time in high-performance jets, would be below five feet eleven in height so they could fit into the spacecraft, be no more than thirty-nine years old and hold a university degree. There was no time or appetite for an open audition. America was racing against the Soviets to claim the scientific, moral and symbolic higher ground. There were no female jet pilots at the time, so that wasn't an option. A list of 508 possible male test pilots were identified from military records. Out of that group, 32 passed the further selection criteria and were dispatched to the infamous Lovelace Clinic in Albuquerque, New Mexico for the most extensive and invasive medical examinations ever performed on the human body.

THE RIGHT CLINIC

William (Randy) Lovelace II, the founder of the clinic, was one of the pioneers of aviation medicine. Randy was a rare breed of human guinea pig, happy to put himself at risk in order to explore how pilots, and astronauts, could survive the ravages of the stratosphere and outer space. In 1943, he jumped out of a B-17 at over 40,000 feet, testing a new high-altitude oxygen supply system. Leaping out of the plane at -45°C (-50°F), and travelling at 200 mph, his parachute opened automatically, the massive jolt rendering him unconscious and ripping off his gloves leaving his hands exposed to the cold. His left hand, whose inner glove was also removed, froze instantly. As well as surviving, he broke the world parachute altitude record in the process. It's worth noting this was the first time he'd ever jumped out of an aeroplane.

Lovelace and his staff designed a week of comprehensive and intimate evaluations for the selected group. But how do you test for the unknown? What are you testing for? And how do you know when you've found it?

THE RIGHT TESTS

Every nook and cranny was scrutinized. Hidden among the impenetrable medical terms, the Lee and Gimlette, the otolaryngological and ballistocardiogram tests, are some of the more eye-watering procedures. There's the use of a dunking stool device, on which the naked candidate was dropped underwater to measure the volume of the body. The prostate examination was especially violent, apparently causing bleeding from the anus: the instrument, once inside, would spread open allowing doctors to peer inside. One candidate referred to the process as 'riding the steel eel'. Cold water was pumped into the ears to test for motion sickness. Jugs were carried by the candidates at all times to pee into, to measure hormone levels, and a succession of barium enemas had to be endured. One of the candidates, John Glenn, commented: 'I didn't know the human body had so many openings to explore.' Even a sperm sample had to be given, making for a memorable scene in the film version of Tom Wolfe's *The Right Stuff*, to observe the effects of radiation on sperm production during space flight.

A less invasive test for you to try is the speech used to check the candidate's vocal clarity, essential for spacecraft radio communication. This passage was used, which contains every vocal sound in the English language. Please read it out loud now:

You wished to know all about my grandfather. Well, he is nearly ninety-three years old. He dresses himself in an ancient black frock coat, usually minus several buttons; yet he still thinks as swiftly as ever. A long, flowing beard clings to his chin, giving those who observe him a pronounced feeling of the utmost respect. When he speaks his voice is just a bit cracked and quivers a trifle. Twice each day he plays skilfully and with zest upon our small organ. Except in the winter when the ooze or snow or ice prevents, he slowly takes a short walk in the open air each day. We have often urged him to walk more and smoke less, but he always answers, 'Banana Oil!' Grandfather likes to be modern in his language.

Below: **Inspecting John Glenn**

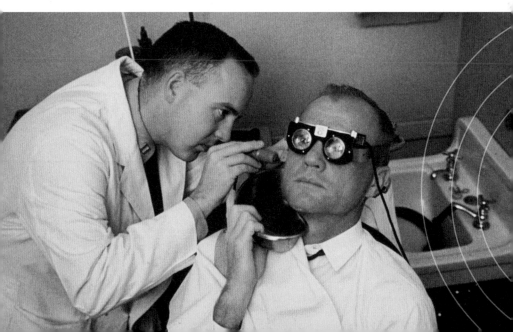

With the medicals done, it was off to Wright-Patterson Air Force Base for physiological stress tests and psychological evaluations. The thirty-one candidates (Jim Lovell had been disqualified) were now assigned letters – A to Z, with the remaining five AA to EE. The physical and physiological tests included:

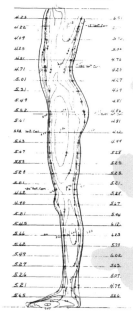

- Tests in the famous centrifuge that spun the candidate around to measure tolerance to the g-force.

- Treadmill and Harvard Step Test (stepping up and down on a step) to measure fitness.

- A sound chamber test in which the candidate would do mental arithmetic while a 145-decibel siren goes off next to his ear.*

- A heat chamber test to evaluate the physiological performance in 54°C (130°F) heat for two hours with a rectal thermometer inserted.

- An anechoic chamber test in a soundproof dark room to see how they would respond to isolation and sensory deprivation.

- A 'complex behaviour simulator' known as the 'idiot box' in which candidates had to respond to signals by pushing the correct buttons and switches.

- The Flack overshoot test. The idea is to hold the mercury column at 40mm by blowing into a tube on a single breath. If it drops below 40mm the clock is stopped.

- The cold pressor test – can you keep your feet in a bucket of iced water for seven minutes? Your blood pressure and pulse taken before/during/after. This was the only test the candidates didn't know about so it would be a complete surprise.

- Equilibrium chair test – the candidate has to maintain their balance on a chair mounted on two hydraulic cylinders, 'counteracting pitch-and-roll disturbances'. Easy enough if you enjoy balancing on your chair at your desk, but try to do it blindfolded.

- Anthropometric studies. Topographic maps were made of each candidate's body giving very detailed measurements. Lines were drawn on their naked bodies to give contrast to the photographs, which were then sent to an aerial photogrammetry company.

* Loud sudden noises and mental arithmetic are my two great fears in life.

As important as it was to be physically healthy specimens, *psychological* profiling was and still is important. The candidates were rated over a number of different categories under the supervision of psychiatrists George Ruff and Edwin Levy. These revealing tests included:

- **Rorschach Test** – By observing the nature of a subject's associations to ten ambiguous ink blots, the psychologist is able to probe relatively deep levels of the personality. Important information on emotions, conflicts and defence mechanisms can be obtained by analysing what is seen and how it is seen.

- **Thematic Apperception Test** – The subject is asked to tell stories suggested by a series of pictures. This test yields information about interpersonal relationships on a fairly deep level.

- **Draw a Person** – By drawing male and female human figures, the subject gives information on his body image and feelings about his place in the world.

- **Sentence Completion Test** – This is a series of incomplete sentences which are completed by the subject. His choice of conclusions provides further personality data.

- **Minnesota Multiphasic Personality Inventory** – An objective paper-and-pencil test which offers a description of the personality based on responses to a 566-item questionnaire.

- **Gordon Personal Profile** – Information on five aspects of personality is obtained by asking the subject to choose, from each of seventeen groups of four statements, the one which describes him best and the one which describes him least.

- **Edwards Personal Preference Schedule** – The subject must choose one statement from each of 225 pairs of self-descriptive statements. This yields scores representing twelve personality dimensions.

- **Shipley Person Inventory** – A test involving twenty pairs of self-descriptive statements related to psychosomatic problems.

- **Outer-Inner Preferences** – The subject chooses one statement from each of fifty-two pairs of statements on feelings about activities, things and other people. This measures interest in and dependency on social groups.

- **Pensacola Z** – By choosing one statement from each of 66 pairs of statements, the subject gives information on 'authoritarian' attitudes.

- **Officer Effectiveness Inventory** – A multiple-choice, self-descriptive test of characteristics related to successful officer performance.

- **'Who Am I?'** – The subject is asked to write twenty answers to the question: 'Who am I?' This is interpreted projectively to give information on identity and perception of social roles.

- **Peer Ratings** – Each candidate is asked to indicate which of the other members of the group accompanying him through the programme he likes best, which one he would like to accompany him on a two-man mission and which one he would assign to the mission if he could not go himself.

The extraordinary comprehensiveness of these physical and psychological tests left nothing to chance. Even the most minor defect was leapt on. From 508 candidates, seven were 'outstanding, without reservations'. According to George Ruff, the chosen ones showed very specific characteristics: '… psychologically healthy men who are realistically oriented to the world, without unattainable ambitions and consequent neurotic concerns, who are adequate to the demands on them and able to function well without emotional distress.'

These seven men were lined up to meet the press like a new boy band. America's first astronauts were: G, K, R, S, U, Z and EE.*

---

* Scott Carpenter, Gordon Cooper, John Glenn, Virgil 'Gus' Grissom, Wally Schirra, Alan Shepard and Donald 'Deke' Slayton.

'SHOULD A GIRL BE FIRST IN SPACE?'

So asks the glorious 1960 American magazine *Look*. Aerobatic aviation hero Betty Skelton is photographed in an iconic silver spacesuit. Glossy *Look* magazine had arranged a photo essay in which Betty would train alongside the new Mercury astronauts. We see Betty smiling her way through the rigorous training procedure. *Look* magazine are quick to point out the serious science behind all of this, as well as imagining the perfect candidate: '… a flat-chested lightweight, under thirty-five years of age and married… her personality will both soothe and stimulate others on her space team'.

One of the sub-headlines reads: 'Experts in our space program predict women will be considered – later.' Those experts were absolutely correct – it would be twenty-three years later before the first American woman

flew in space. A flick through the rest of the magazine is illuminating as to at least one reason why. A snapshot of middle-class America in 1960. The adverts remind us just how rigidly defined gender roles were: pristine women in domestic and caring roles surrounded by young children. Men are either shown at work (proper work of course), or being confronted with that other mid-century masculine American fantasy, the Marlboro man cowboy. The space-cowboy.

As the idea of women astronauts began to circulate in the media, as a curiosity more than anything else, new feminized space names were being dreamed up: 'Astronautrix', 'feminaut', 'astronette'. But that's not to say the idea of women in space wasn't being taken seriously by some. Randy Lovelace had devised a project to look at the question of comparing women against

AŤ ŽIJE
PRVNÍ KOSMONAUTKA
SVĚTA!

men in terms of their suitability for space travel. Signs were showing that women had various physiological advantages – fewer heart attacks, as well as the physical benefits of being smaller and lighter – and were outperforming men in a variety of studies such as isolation and sensory deprivation. It was Jerrie Cobb, a record-breaking female aviator, who was approached to be the guinea pig and became the first woman to take the identical medical examination as the official Project Mercury men,* passing 'beyond expectations'. There were, however, plenty of astounding female aviators who might have the 'right stuff'. In all, twenty-five women completed the Lovelace tests, with thirteen passing, who became known as the First Lady Astronaut Trainees (FLATs) or the Mercury 13.

To date, 60 out of the 553 astronauts have been women. The latest batch of eight NASA astronaut candidates is a 50:50 gender split; the Soviets sent up Valentina Tereshkova as early as 1963 on Vostok 6; and the UK's first astronaut was Helen Sharman. Sally Ride finally got American women into space in 1983 on STS-7 as a mission specialist, but it was Eileen Collins who became the first astronaut 'pilot', something that was celebrated. It was the Shuttle that opened space travel up for everyone.** The Project Mercury capsule could carry one person, Apollo three, but the Space Shuttle was able to fly much bigger crews (seven) which meant that the selection pool could be opened much wider, broadening the role of the astronaut.

---

* Minus the sperm and steel eel, plus some gynaecological tests.

** Not everyone, obviously.

AM I AN ASTRONAUT?

CANDIDATE NAME:	YES	NO
Bachelor's degree in engineering, biological science, physical science, computer science or mathematics*		
At least 3 years of related, progressively responsible professional experience** OR At least 1000 hours of pilot-in-command time in jet aircraft		
Distant and near visual acuity correctable to 20/20 in each eye		
Blood pressure not exceeding 140/90 measured in a sitting position		
Height between 62 and 75 inches		
Official SCUBA qualification		
Swimming 3 lengths of a 25-metre pool without stopping. THEN swimming another 3 lengths of the pool. In a flight suit. And wearing tennis shoes.		
Treading water continuously for 10 minutes wearing a flight suit		
Вы говорите по русски?		
AM I AN ASTRONAUT?		

* The following degree fields are not considered qualifying: technology (engineering technology, aviation technology, medical technology, etc); psychology (except for clinical psychology, physiological psychology or experimental psychology); nursing; exercise physiology or similar fields; social sciences (geography, anthropology, archaeology, etc); and aviation, aviation management or similar fields.

** An advanced degree is desirable and may be substituted for experience as follows: master's degree = one year of experience, doctoral degree = three years of experience. Teaching experience is considered to be qualifying experience.

Opposite: **Valentina Tereshkova**

TRAVEL GUIDE: THE STEELY EYED MISSILE MAN

Name:
John Aaron
Profession:
EECOM Apollo 12
Claim to fame:
Being in the right place at the right time

On 14 November 1969, four months after the first moon landing, the Apollo 12 Saturn V rocket lifted off once again into the thick blanket clouds of the Florida sky at 11.22 EST. On board were Alan Bean, Pete Conrad and Dick Gordon. One minute after a perfect lift-off, something happened.

Unknown to anyone, the problem had been caused when lightning had struck the rocket as it passed through the clouds. Apollo astronaut Jim Lovell later reflected on the gravity of the situation: '... the flight rules dictated an abort. When six million pounds of fully fuelled, freshly launched Saturn 5 begins flying out of control, you don't wait for engineering analysts to tell you what's gone wrong. You light the escape rockets at the tip of the booster, accelerate the capsule away from the Saturn and blow the whole wayward missile over the empty Atlantic.'

That's one option. But they were lucky that John Aaron was on duty in mission control that day. He was the EECOM

CONFIDENTIAL

00 00 00 38	CDR	Huh?
00 00 00 39	CMP	I lost a whole bunch of stuff; I don't know - -
00 00 00 40	CDR	Turn off the buses.
00 00 00 41	CC	MARK.
00 00 00 42	CC	One Bravo.
00 00 00 43	CDR	Roger. We had a whole bunch of buses drop out.
00 00 00 45	LMP	There's nothing - it's nothing - -
00 00 00 47	CMP	A circuit -
00 00 00 48	CDR	Where are we going?
00 00 00 50	CMP	I can't see; there's something wrong.
00 00 00 51	CDR	AC BUS 1 light, all the fuel cells - -
00 00 00 52	MS	...
00 00 00 56	CDR	I just lost the platform.
00 00 00 59	CMP	All we've got's the GDC.
00 00 01 01	CDR	Yes. Okay. We just lost the platform, gang. I don't know what happened here; we had everything in the world drop out.
00 00 01 09	CMP	I can't - There's nothing I can tell is wrong, Pete.
00 00 01 11	CDR	I got three FUEL CELL lights, an AC BUS light, a FUEL CELL DISCONNECT, AC BUS OVERLOAD 1 and 2, MAIN BUS A and B out.
00 00 01 21	LMP	I got ac.
00 00 01 22	CDR	We got ac?
00 00 01 23	LMP	Yes.
00 00 01 24	CDR	Maybe it's just the indicator. What do you got on the main bus?
00 00 01 26	LMP	Main bus is - The volt indicated is 24 volts.

CONFIDENTIAL

(Electrical, Environmental and Consumables Manager) flight controller in charge of overseeing the electrical systems, whose job it was to report to the flight director Gerry Griffin. In front of him on the screen the telemetry data from the spacecraft, which moments ago were showing perfect readings, suddenly went wild throwing up patterns of seemingly random numbers. The Command Module had lost power and was now running on emergency batteries, and the resulting warning lights and alarms sent the astronauts and Mission Control into confusion. A split decision had to be made – should they proceed? Or abort and blast the crew to safety and detonate the rocket before it's too late?

John Aaron was only twenty-six years old. His mother was a minister and his father a cattle farmer. Studying maths and physics, he planned to become a teacher after leaving college like most of his seven sisters, but on the advice of a friend sent off a speculative application to NASA. Without even having to sit an interview he got the job, cutting his teeth on the Gemini programme. He was blessed with a deep curiosity about how things worked, wanting to understand, not just his own specific area of expertise, but how the myriad different systems on a space mission worked together.

Facing a potentially catastrophic situation they had only moments to make the call. Then, for John Aaron, the penny dropped. By pure chance, he recognized this exact same pattern of random-looking telemetry numbers from a year earlier during a practice simulator session. He calmly addressed Gerry Griffin: 'Flight, try SCE to AUX…' No one on the ground or in the air knew what he was talking about. Inside the Command Module Pete

Conrad replied, 'FCE to AUXILIARY? What the hell is that?' SCE (signal conditioning equipment) to AUX (the auxiliary position) was an obscure switch on the main Command Module instrument panel that everyone had forgotten about. Everyone except John, who remembered the switch fixing the problem during the simulator run. On board the switch was located by Alan Bean, and lo and behold the telemetry data went to back to normal. John's calm manner and holistic approach to understanding the vehicle's complex systems saved the day, earning him the moniker as NASA's original 'steely-eyed missile man', a sentiment that sums up what goes on in the mind in that elusive gap between a stimulus and a response. John was lucky. But luck is something that visits more frequently to those who come prepared.

I witnessed my own SCE to AUX moment at the Royal Albert Hall in London recently with Tim Peake and Tim Kopra. Moments before we were due to walk on stage to give a talk, there was a problem (as there always is) with the PowerPoint presentation. The usual scene of flustered people prodding a laptop and scratching their heads ensued. Tim Peake immediately snapped into astronaut mode, taking control of the situation, 'working the problem' and getting it calmly and efficiently sorted.

Whether it's a PowerPoint presentation gone rogue, or a freshly fuelled Saturn 5 being hit by lightning, there are some people who just have the right stuff.

YOU'VE GOT TEN SECONDS TO FIND THE SCE TO AUX SWITCH ON THE COMMAND MODULE INSTRUMENT PANEL BEFORE THIS BOOK SELF DESTRUCTS. GO:

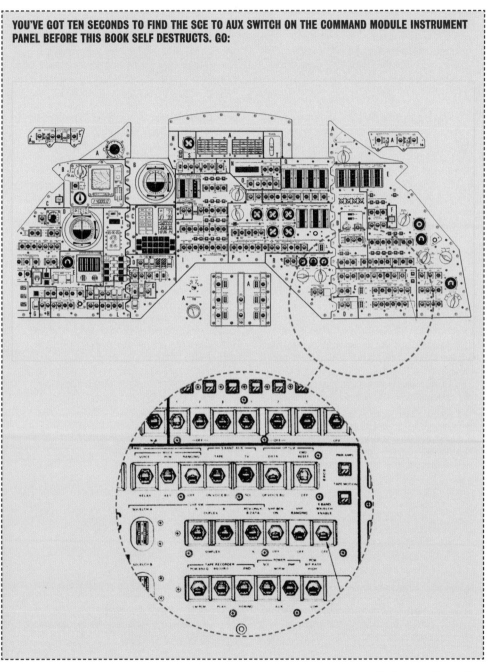

THE AFRONAUTS

'Zambians are inferior to no men in science and technology. My space plan will surely be carried out.'
Edward Nkoloso

'We're going to Mars! With a spacegirl, two cats and a missionary,' proclaimed Edward Mukuka Nkoloso, the revolutionary, patriot, teacher, philosopher, scientist and dreamer. Like a moth drawn to the light, he wanted not only to escape the bonds of colonial rule, but also the shackles of earth's gravity. Nkoloso wanted Zambia to touch the stars. So he did what any political freedom-fighter-cum-space-traveller would do – created his own national space agency.

The National Academy of Science, Space Research and Philosophy was founded in his home city of Lusaka in the grounds of an abandoned but elegant farmhouse in 1964, just a few years after Gagarin first left the cradle. It even had a motto designed to inspire and distil the essence of their values: 'Where fate and human glory lead, we are always there.' Eleven men were recruited to his astronaut corps, as well as a teenage girl called Matha Mwambwa. Like America's first group of seven astronauts, Nkoloso's selected group were put through a series of rigorous physical training exercises: a mixture of jumping jacks, callanetics and sit-ups, as well as g-force and weightlessness training, which involved playground swings and rolling down a hill in an oil drum – if you emerged unable to stand, you were disqualified. Astronomy and orbital mechanics were taught outside, where the recruits would learn about the stars and the moon. Nkoloso began imagining rocket and capsule designs, and in an act of early international cooperation he approached both the Russians and Americans to propose collaboration: a Zambian and an American should walk on the moon together, with the Zambian flag first to go up. That was a deal-breaker. He was convinced that Mars was inhabited 'by primitive natives', and proposed a launch to coincide with Zambian Independence Day. His request was turned down, but a rocket, the D-Kalu 1, named after President Kaunda, was made from some oil drums welded together. In a TV interview, we see Nkoloso dressed in a flamboyant cape and tin helmet, with his ragtag army of guerrilla astronauts training behind him, demonstrating to the world Zambia's fitness and intention. Not just political independence, but planetary independence too. Not even gravity could clip their wings. The interviewer addresses us, unsure as to whether he is witnessing a hoax or madness.

Like all dreams, the Zambian space programme began to fade. The space girl became pregnant, the Afronauts drifted into employment elsewhere, and Nkoloso himself went back into mainstream politics, running for mayor and eventually studying law. The story to this day lives on through artists and film-makers, most vividly by the artist Cristina de Middel, who reimagined this story creating African-inspired spacesuits, juxtaposed with landscapes.

Behind the Nkoloso story lies neither a hoax nor madness. It is a moment that expresses the hopes of a country, the future of mankind, and indeed why we want to leave the planet at all.

'Look at that tree. Because I can see the tree I can go to the tree. It's the same with the moon.'
Edward Nkoloso

Opposite: **Hamba from the series The Afronauts, Cristina De Middel (2012).**

'It is an ordinary man that will make this extraordinary journey. And ordinary men will make it possible. They're making him a suit for strolling out on the moon. Something special. Cut not from fine woollens but from the best grade aluminum. The latest style in neoprene rubber, elastic web, plastic tubing, nylon, cheese cloth. Six layers in all, with pockets extra big for picking up moon stones... And here too, the months and years of planning, testing, experimenting, accepting this concept of design, rejecting that hypothesis... here too the long labour comes down to something you can touch. Something which can be glued, and sewed, and taped...'
NASA archive film about ILC Dover

One of the most famous photographs ever taken is of Buzz Aldrin standing in the Sea of Tranquillity. But we don't really see 'Buzz Aldrin' at all. What we're actually looking at is an engineered object. A piece of space hardware. At the time, the most advanced, expensive, technically challenging item of clothing ever devised and made. This is the A7-L* lunar excursion suit, made by ILC Dover: a division of the International Latex Corporation, later known as Playtex, best known for manufacturing women's structured underwear. Perhaps the most iconic symbol of human endeavour, the most instantly recognizable suit of clothes ever made. The word suit doesn't really do it justice – this is a wearable spacecraft, engineered to protect the 'ordinary man' from the extremes of the lunar surface. Twenty-one layers of new exotic materials engineered by the finest minds and stitched together by a production line of highly skilled seamstresses from Delaware, who before this had been stitching suitcases and boxing gloves: heroes of engineering and technology whose names are sadly overlooked, such as Michelle Tice, Julia Brown, Delema Austin, Delores Zeroles, Doris Boisey and Delema Comegys. The A7-L and Hamilton Standard life-support backpack that went with it became part of one of the defining images of the twentieth century.

SUITS OF ARMOUR
The complexity of the A7-L didn't happen in a single evolutionary leap, but represents the entire history of human exploration, designed to protect the fragile human body from hostile forces. We can see its ancestors in the metal exoskeleton suits of armour, designed to withstand impact in battle. Then there are the Victorian diving suits with their brass helmets, lead boots and air hoses. And the woollen balaclavas of the polar explorers, with their wooden goggles with horizontal slits to protect against the cold and the

---

* A = Apollo, 7 = the number in the series, L = the 'L' from the company name.

brightness of the reflected sunlight on the ice. There are the dry suits and helmets of industrial divers, and the pressure suits and g-suits of pilots, all of which were engineered to take men and women to realms normally too extreme for humans to function.

A spacesuit must protect the wearer from any lack of atmospheric pressure and provide breathable air and remove carbon dioxide. It has to insulate against extreme variations in temperature and be strong enough to shield the body from the impacts of micrometeoroids – tiny pieces of space rock or dust whizzing about like bullets. It must also protect from radiation and sunlight glare, and be tough enough to withstand whatever abrasion comes its way. Our imaginations have evolved for space travel. Our bodies have some catching up to do.

UNDER PRESSURE

Down here on planet earth, natural selection has equipped us to live quite happily at the bottom of an ocean of air we call our atmosphere which, along with the earth's magnetic field, protects us from much of what the universe throws at us. That atmosphere has weight, which we call air pressure. If you're at sea level, hold your hand out and draw a one-inch square on your palm. Within that square you are holding a column of air that stretches from your hand up to space that weighs 14.7 pounds (psi or pounds per square inch). The higher you go, the thinner the air becomes and the pressure (weight of air) decreases as a result. If you go too high, you'll run into problems.

The Armstrong Line[*] at 63,000 feet (19 km) is a demarcation line above which atmospheric pressure is so low that it'd be impossible to survive without a pressure

ATMOSPHERIC DATA
for a 'standard day' (59°F/15°C) at 40 degrees latitude

Altitude (feet)	Atmospheric Pressure (psi)	Temperature (°F)	Temperature (°C)	Time of Useful Consciousness
100,000	0.15	-51	-46	0
90,000	0.25	-56	-49	0
75,000	0.50	-65	-54	0
63,000	0.73	-67	-55	0
50,000	1.69	-67	-55	0-5 seconds
43,000	2.40	-67	-55	5-10 seconds
40,000	2.72	-67	-55	10-20 seconds
35,000	3.50	-66	-54	30-60 seconds
30,000	4.36	-48	-44	1-3 minutes
25,000	5.45	-30	-34	3-5 minutes
20,000	6.75	-12	-24	10-20 minutes
18,000	7.34	-5	-21	20-30 minutes
15,000	8.30	5	-15	30+ minutes
10,000	10.11	23	-5	Nearly Indefinitely
7000	11.30	34	1	Indefinitely
Sea Level	14.69	59	15	Indefinitely

* Named after the aeromedical pioneer Harry George Armstrong.

suit. Up here water boils at 37°C (body temperature). The pressure is so low that your body's fluids would simply boil away.*

Pilot William Rankin knows what that feels like, in what must rank as one of the worst days ever, and a salutary lesson in why you should always dress for your environment. At over 45,000 feet he was forced to eject without a pressure suit through the canopy of the pressurized cabin of his F-8U, exposing himself to an explosive decompression, and a sudden drop in temperature from 24°C to around -55°C in an instant. He vividly describes what the exposure to such a low pressure environment feels like:

'It was nature's cruellest torture, the screw and rack of space, the body crusher, the body stretcher, each second another turn of the screw, another wrench of the rack, another interminable shot of pain. Once I caught a horrified glimpse of my stomach, swollen as though I were in well advanced pregnancy. I had never known such savage pain. I was convinced I would not survive; no human could...'

Bleeding from his nose, ears and mouth he managed to survive the ordeal, despite on top of everything else ending up free-falling through a lightning storm.

Such explosive depressurization sealed the fate of the Russian crew of a Soyuz spacecraft in 1971, killing three cosmonauts who were asphyxiated in the descent module while returning to earth from the space station Salyut 1. They weren't wearing spacesuits at all. Since that accident, all astronauts who fly on Soyuz wear the white with blue trim Sokol spacesuit – an emergency pressure suit – to prevent just such gruesome catastrophes.

Above: **Allyn B. Hazard's Lunar Exploration Suit Mark 1**

---

* Just the exposed fluids like saliva and tears, not your blood.

Hypoxia, ebullism, anoxia – the list of bad things that will happen to you with exposure to low-pressure environments are well known to astronauts, flight surgeons and enthusiastic children with a morbid curiosity the world over. Sadly, the eye-bulging demise of Arnold Schwarzenegger in *Total Recall* on the Martian surface embellishes a popular myth: you won't explode in the vacuum of space or on the low pressure Martian surface. Your strong stretchy skin and tissues will hold you all together. If you found yourself thrown out of a spacecraft, trying to hold your breath might seem like a smart move, but it will only rupture your lungs. The outlook is bad. Your death would be silent. There would be no last words. In the vacuum, you've got about ten seconds of consciousness to ponder the situation. Enough time to think a profound final thought. Something like 'It's all full of stars' would be all you'd have time for, before the lights went out. And then in a minute or so you'd be gently smothered by a silent universe of chaos and indifference.

SPACESUIT HISTORY

The story of the spacesuit begins in earnest with the birth of aviation. For early pilots in open cockpits it was all about keeping warm, but new solutions were required the higher we climbed. Maverick pilot Wiley Post was famous for losing an eye (oil field accident) and a stint in prison (stealing a car), as well as becoming the first pilot to fly around the world solo in 1933 in his Lockheed Vega the *Winnie Mae*.

Not one to rest on his laurels, Post understood that he could fly even faster if he could access the jet stream – currents of fast-moving air which occur at around 50,000 feet. He was convinced that the future of aviation was up in the stratosphere, but was also well aware of the physiological problems involved in surviving there. Constructing a fully pressurized cockpit for his aircraft would be too heavy, complicated and expensive, and so

Opposite: **Aviator Wiley Post**

he devised a solution: 'My idea is to employ a suit, something like a diver's outfit, which can be blown up with air or oxygen to the required pressure.'

Post sought the help of the B. F. Goodrich Company, famous for manufacturing car tyres. There he teamed up with Russell S. Colley, who had been interested in women's fashion design, but was persuaded into an engineering career. He would become known as 'the first tailor of the space age'.

In time-honoured tradition it took three attempts to get it right. The first two suits failed – the first one splitting apart under pressure, and for the second one Post had gained weight and had to be cut out of it. The third suit fitted perfectly and was pressurized enough to keep Post alive, while at the same time allowing him to maintain flexibility to control the aircraft. Too much air pumped

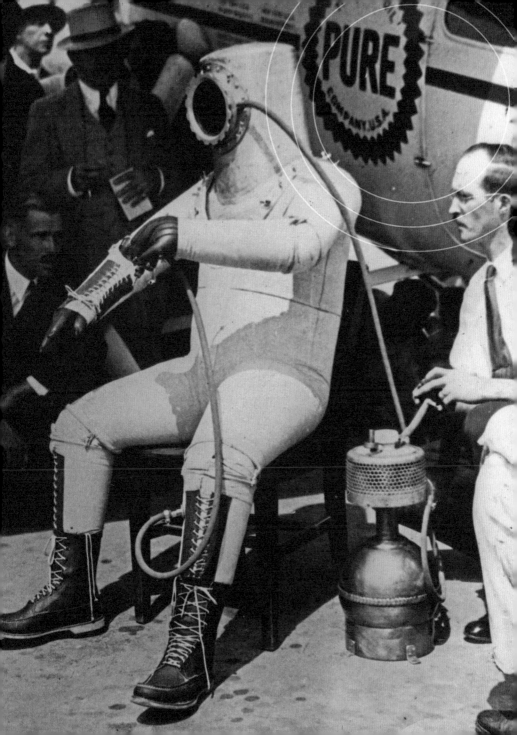

into a suit will make it rigid – the more air you blow into a long modelling balloon, for example, the stiffer it becomes. Likewise with a spacesuit – over-inflation means the wearer has to put in more work to move around. It's this trade-off between pressure versus flexibility that would become such an important engineering challenge to spacesuit design.

Seeing the component parts of what is essentially the very first spacesuit, you understand the basic design philosophy that underlines all the suits that followed: a cotton onesie for warmth and comfort, the strange-looking airtight rubberized inflatable bladder that looks like a human-shaped bicycle inner tube, the heavy canvas outer suit to stop the bladder ballooning, which has articulated joints, leather gloves, boots,

and the aluminium helmet looking like a tin can with a porthole set slight off centre to compensate for Post's missing eye. It's a strangely disconcerting Heath-Robinson-esque outfit, straight out of an early science-fiction comic strip.

THE XH-5 TOMATO WORM SUIT
The many strange looking prototype suits that were to follow solved many engineering challenges, which in turn fed the imagination of post-war science-fiction writers and artists. The XH-5 B. F. Goodrich pressure suit of 1943 is a classic example. A major design innovation by Colley was inspired by watching a tomato worm caterpillar crawling along a branch, or so the story goes. How does a caterpillar bend its body without squashing itself? Colley noticed that as it moved along,

Above: **Wiley Post's pressure suit**
Opposite: **Tintin, Snowy and the B.F. Goodrich 'tomato worm' suit**

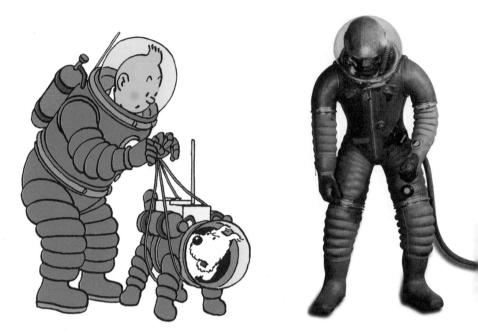

its segmented body would contract at the bottom while expanding at the top, thus maintaining its internal volume. Imagine for a moment a drinking straw: if you bend it in half, creating a kink, the straw becomes useless. But a clever engineer* devised that bendy concertina section that allows the straw to flex, maintaining the internal volume just like the caterpillar. In spacesuit engineering, these bendy jointed sections are called *convolutes*, partially solving the problem of maintaining suit flexibility under pressure, whilst allowing astronauts to work.

Although Colley's 'tomato worm' suit never flew in space, it did make it to the moon in Tintin's *Destination Moon* in spectacular orange. It also came in a handy dog version.

New exotic materials were being considered for pressure suits, like Dacron, Mylar, Nomex, neoprene rubber, Kapton,

Chromel-R and Beta cloth. The experimental X planes that would fly at hypersonic speeds at the 'edge of space' in the 1950s and 1960s required ever-more sophisticated outfits for the pilots. The young X-15 pilot Neil Armstrong wore the latest pressure suit made by the David Clark Company, which was to set the tone. What made it even more exciting was that it was silver.

SILVER SUITS

It was the Project Mercury suit design that really ushered in the space age. A suit that met all the technical challenges of space travel, as well as igniting the public imagination. Like Buck Rogers – commander of earth's interplanetary battle fleets – they would be silver. Once again it was designed by Russell Colley and the B. F. Goodrich Company. The 'aluminized' silver colour came from an

---

* Joseph B. Friedman

aluminium powder that was bonded onto the green nylon of the Navy Mark IV pressure suit, and was made by 3M (Minnesota, Mining and Manufacturing Company), which you probably know as the company that make that other engineered marvel of the twentieth century, the Post-it note.

The technical justification for the silver suit was UV reflection and various thermal properties, but the *aesthetics* are what's important here. Silver elevated the astronaut to the status of a cultural icon. Ask anyone who's wrapped themselves in acres of kitchen tinfoil as a child to create the same effect. They look beautiful, tailored and sleek. If you want a vintage spacesuit in your house, you want one of these.

If you look back at the famous Mercury 7 picture, you'll notice that Deke Slayton and John Glenn's boots are slightly different. Their custom-made boots weren't ready in time for the famous *Life* magazine photoshoot, so they had to spray-paint a pair of ordinary black workboots silver.

SOKOL

For the last thirty years, the real workhorse spacesuit has been the Sokol ('Falcon') suit made by NPP Zvezda in Moscow. Every astronaut (including you) who flies in the Soyuz spacecraft wears one. It's a soft nylon suit with integrated boots, hood and visor and a rubberized internal pressure bladder. It's designed to be worn in a reclining position in the specially moulded seat, which is why they look a bit small when astronauts are walking in them. The glorious bit is how you seal the pressure bladder, which looks like a sort of thin rubberized sleeping bag that spills out of the front, and is what you climb into to put the suit on. Once you're in, the opening is then pleated up and tied off with nothing more sophisticated than a rubber band. Like a pencil, a rubber band is cheap, and if it breaks, you can just grab another one.

With the gloves on and the visor closed, the suit can be inflated through the hose on the front just like a human-shaped balloon, protecting the astronaut from emergency depressurization.

EVA

On board the ISS there are two types of spacesuit for going outside: the rear-entry Orlan suit used by the Russians, and the American EMU (Extravehicular Mobility Unit) version made by ILC Dover, which are used by everyone else. There are two of these EMU suits permanently on board, which all the visiting astronauts use. In 2013, ESA Astronaut Luca Parmitano had a major issue when water from the cooling unit leaked inside his helmet while he was outside the station on an EVA.[*] You really don't want this to happen. Luckily he was quickly back inside before the problem developed further.

There are wardrobe malfunctions, and then there's the wardrobe malfunction of the first space walker Alexei Leonov. On 18 March 1965, he and cosmonaut Pavel Belyayev made it into orbit aboard Voskhod 2. Already, conditions on board were fearsome: overheating and claustrophobia, combined with motion sickness from the effect of weightlessness, had already set in. Any body movements in such a tiny space were almost impossible. Both were wearing the new Berkut spacesuit – technically similar to Gagarin's orange SK-1 pressure suit, but made from a white outer fabric designed to reflect the solar radiation and finished with a smart red stripe down the leg and arm like a Soviet general's uniform, providing a visual contrast.

It was time for Leonov to finally open the door and go outside, the first time any human had done so. In a spacecraft no bigger than a phone box, Leonov began the exhausting exit procedures. Like everything about this mission, how this new-fangled spacesuit would perform was unknown. It had been

---

[*] Extra-vehicular activity.

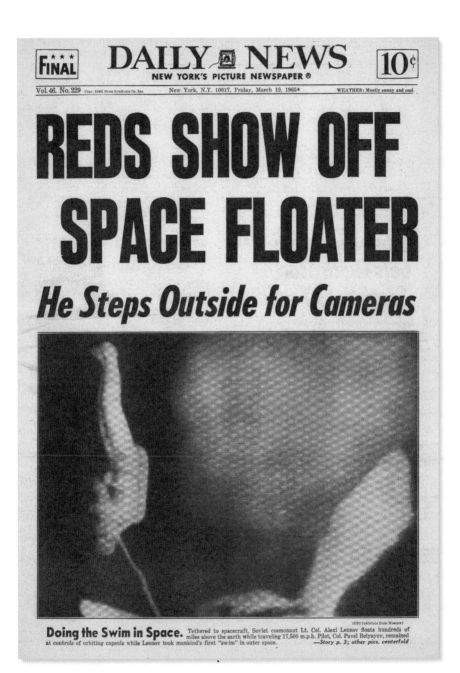

tested on earth but never in such an extreme environment. Belyayev deployed the airlock – a soft extendable tunnel that protruded outwards from the spacecraft. A corridor into the void joining two realms. With a slap on his back he sent Leonov on his way, who closed the hatch behind him, then slowly released the air from this flimsy airlock before opening the hatch at the other end. Leonov made his way to the exit rim almost unable to move in his cumbersome inflated suit and set up the movie camera, which was to record this historic moment. The whole of Africa lit up in front of him, filling his field of view. Breathing pure oxygen, he was attached by a five-metre long life-support line to the spacecraft.

If ever there was a metaphor for the birth of the space age this was it: Leonov free-floating like a foetus wrapped in a protective cocoon, tethered by an umbilical cord to the womb. 'Caucasus! Caucasus! I see the Caucasus underneath!' he declares over the radio, already exhausted from the exertion of trying to move in the suit. The countries of the world pass below him like a scrolling map. To his left the boot of Italy, and to his right the Black Sea. He hears nothing but his breathing and his beating heart. Like a newborn his movements are small and deliberate, conserving as much energy as possible, fighting against the pressure of the inflated suit.

Moving his arm to take a photograph using the camera attached to his chest he realizes he is in trouble. The suit had expanded too much, causing it to stiffen like an over-inflated balloon. This forces the gloves and boots away from his hands and feet. With nothing to push against in the vacuum, and unable to hold the tether, he is stranded, despite being only a couple of metres from the airlock.

The danger he's in begins to dawn on him. He knows in five minutes the earth will move into shadow, which means total blackness as well as mind-numbing cold. Leonov begins frantically trying to get back to the airlock. The bulging suit is no longer tight around his body, making any movement useless. His gloveless fingers can no longer grasp. He tries to clamp his arms around the tether to pull himself nearer. Every movement is energy-sapping and generating heat. Sweat is pouring into his eyes, blinding him.

Eventually, managing to move towards to the airlock, he realizes he is physically too big to get back in. It is as futile as trying to get the toothpaste back into the tube. He has no choice but to release half the pressure of the suit using a manual valve, which he does in secret. Reducing the pressure so drastically might give him more flexibility and reduce his size, but he knows it could also kill him. Sure enough as Leonov struggled with the valve, and the oxygen seeped slowly out of the suit, he could feel death's first gentle embrace – the telltale pins and needles creeping over his fingertips. But his decision does give him the increased flexibility he needs to get himself to safety, and head first he manages to force himself back inside the spacecraft…

He was suffering from exhaustion and dehydration and had lost 6kg of body weight in half an hour. But he was alive and ready to return to earth, a journey that almost ended in disaster: a spacecraft separation issue meant a ballistic re-entry, entering the atmosphere at too steep an angle exposing the cosmonauts to over 9-g and sending them wildly off course. They landed deep in the frozen wilderness of the Soviet Union and had to survive for several days waiting for rescue. Luck was on their side: they had survival training and a gun, which meant they could kill any of the wolves and bears living in the forests. But that's another story.

RED PAINT

One of the defining visual images of the Soviet space program was the red CCCP (Cyrillic for USSR) filling the expanse of white on the helmet of Yuri Gagarin. This historic paint job was a last-minute decision, just a few hours before launch, as much for identification on landing as anything else. A Zvezda engineer named Victor Davidyantz was the man tasked with this important afterthought. Working against the clock, with a steady hand and nerves of steel, he inadvertently created a design classic that even got a feature in a William Hartnell Doctor Who episode. If you go to the Smithsonian Museum in Washington DC, you can see one of Gagarin's training suits and the signwriting handiwork up close. The red paint has been chipped away slightly, but the brushstrokes of the artist are clearly visible. The letters are ever so slightly varying in thickness, and as with any artefact where you can see the hand at work, you are instantly transported back in time. The red also provided good visual contrast for the cameras. Later versions of the famous Apollo moon suit, worn by the Commander during lunar EVAs, had a red stripe on the helmet and limbs for easier identification.

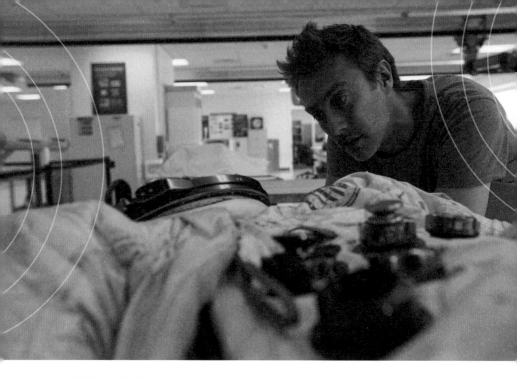

A BETTER MAN THAN ME

In 2015, a Kickstarter campaign 'Reboot the Suit' was set up to fund the restoration of Neil Armstrong's historic moon suit, which is badly degrading after nearly fifty years. Nearly three quarters of a million dollars were raised almost immediately, so high is the spacesuit held in the public's esteem. The spacesuits of the past have huge cultural importance, as well as holding valuable information while we consider what to wear as we journey further from earth. The majority of the American suits from the Mercury, Gemini and Apollo era are now looked after by the National Air and Space Museum at the Smithsonian. Going into the large pristine laboratory where some of the suit preservation and imaging work is being carried out is like walking into a morgue – human forms lie on metal gurneys

covered by cotton shrouds. Under one is Al Shepard's Project Mercury B. F. Goodrich suit, the aluminized silver coating beginning to fade and fall away.

In front of me like a corpse lies Charlie Duke's Apollo 16 suit. An open zip reveals its various layers, like looking into the wound on the carcass of an animal. Every stitch and every stain tells a story, that began with a group of highly skilled women in the sewing rooms in Delaware, all the way to the mountains of Fra Mauro on the moon and back. I notice that the blue lining, just inside the red helmet neck ring, has discoloured – a mysterious darker patch, as if he'd been drooling. What could it be? A couple of years later, I come across the name 'Gunga Din' – the title of a Rudyard Kipling poem about an Indian water-bearer who saves the life

Above: **Author with Charlie Duke's A7LB spacesuit**
Opposite: **Buzz Aldrin's name tag**

of an English soldier. Gunga Din was also the nickname the suit engineers gave for a specifically designed drinking pouch that attached right by that spot in the suit. It turns out that Tang, the soft drink that Charlie had in this pouch, had leaked out during one of his lunar EVAs, causing the helmet locking mechanism to jam.

Through the microscope we peer back in time, deep into the texture of the fabric. The weave of the individual fibres stained grey from the moon dust, with grit embedded between them like boulders strewn across a landscape. Some of the fibres had snapped or been damaged by the abrasions. These suits were engineered to withstand everything that the universe could throw at them. Everything except time. The damage done by a few hours of wear and tear on the moon and a few decades of hibernation on earth has taken its toll, but gives us vital knowledge for the next generation of spacesuit designers.

> Tho' I've belted you and flayed you,
> By the livin' Gawd that made you,
> You're a better man than I am, Gunga Din!
> Rudyard Kipling

SPACESUITS, THE NEXT GENERATION

'She will not be bosomy because of the problems of designing pressure suits'
Look magazine, 2 February 1960, on what the first 'space girl' might be like

The next generation of spacesuits for the next generation of spacecraft are already being rolled out. They've got to be functional, practical, and importantly they've got to look bad-ass. Spacesuits sell the dream of space travel. The David Clark Company has manufactured the new 'Boeing Blue' suits, made for the Starliner spacecraft. Like the Soyuz's white Sokol suit and the Space Shuttle's pumpkin-orange Advanced Crew Escape Suit (ACES), this suit is designed for on-board emergencies during the critical launch and land phases. This new suit is much lighter and more flexible, with touchscreen sensitive gloves, an extra wide visor, and a soft integrated helmet like the Sokol.

NASA's new Orion spacecraft will use a David Clark Company orange ACES suit, similar to the Space Shuttle suit, but modified (MACES) with a limited degree of EVA capability.

ESA GRAVITY LOADING COUNTERMEASURE SKINSUIT (GLCS)

Microgravity environments like the ISS, or year-long trips to Mars, will lead, among other things, to muscle wastage and mineral bone density loss. Without your body naturally working against gravity, your whole physiology will quickly deteriorate. The Skin Suit, which has already been tested on board the ISS, squeezes the body from the shoulders to the feet while the astronaut carries out their duties, to help overcome some of the damaging effects, specifically helping to prevent back pain caused by the lack of compression on the spine, which can make astronauts grow by a few inches during a long duration spaceflight.

EXTRAVEHICULAR

There's no such thing as bad weather in space. Just the wrong spacesuit.

The suit you will wear to explore your planetary destination will have to be much more practical than anything built before. If you're planning to be on Mars for a year in your spacesuit you'll need vastly increased mobility and durability. It will have to last a long time – the Apollo A7-L moon suits were only used once, and for a matter of hours or days. The gravity of the planet will determine the weight of the suit. The current NASA EMU (Extravehicular Mobility Unit) suit used for spacewalks on the ISS weighs nothing at all up there, but a whopping 140 kg here on earth. Many new solutions are needed depending on where we're going, for how long and what we're going to do.

Professor Dava Newman, an astronautics engineer, and former NASA Deputy Administrator has been developing the

Above: **NASA astronaut Sunita Williams in the Boeing Starliner spacesuit**

BioSuit, a futuristic-looking Spiderman suit that works by using mechanical counter-pressure (a physical squeezing of the body) rather than the traditional 'balloon-style' gas-pressure suits. A suit like this would give the wearer a huge degree of flexibility. The idea is not new and was explored in the 1970s, but now a new world of materials and design techniques is making it possible.

ILC Dover in Delaware continue the gas-pressurized spacesuit theme, designing the next generation of lunar and Mars suits, using rigid, soft and hybrid designs, such as the prototype Z-2 suit.

If you've always wanted to put on a real spacesuit, you can try buying one (or bits of one) on eBay. Periodically Russian Sokol suits and Orlan EVA gloves appear from mysterious sources online, or occasionally they come up via more reputable auction houses. Even so, you're talking about tens of thousands of dollars.

If you're in Brooklyn, check out independent spacesuit manufacturers Final Frontier Design. It's run by Ted Southern and former Zvezda Russian spacesuit engineer Nikolay Moiseev who set out on a mission to try to make a better spacesuit glove – traditionally one of the hardest components to get right. They have now set up a spacesuit design company to cater for the new world of space access. You can spend a couple of hours with them, going through a real suit-up procedure for $795.00.

Above: **The form-fitting Biosuit**

Above: **Z-2 prototype**

DO I NEED A VISA?

If you're planning on a space trip, make sure you don't get caught out by red tape and bureaucracy. Dr Jill Stuart, from the London School of Economics, an expert in space law, gives us some advice about the necessary paperwork.

1 Do I need a passport to go into space?

Not yet. Space remains the domain of so few people that it's straightforward keeping track of who has been up there. You also don't need the added weight of identity papers. But this doesn't mean you'll be anonymous. All your details will be logged with the domestic government of the country you are launching from, as well as the United Nations and other space-monitoring bodies. Governments in particular would be legally responsible if something happens to you, so they will be watching you carefully.

Outer space law was written when astronauts were few, highly vetted and government-employed. With the rise of private companies offering to take civilians into space, an international space passport regime may well be on the cards. If you crash back to earth you are considered to be an 'envoy of all mankind', and whatever country rescues you must be nice to you and send you home. Your home government would likely get sent a bill for the recovery effort.

2 Do I need a visa to land on another planet?

No. According to outer space law, all 'celestial bodies' in the universe are considered to be 'neutral territory', not subject to the appropriation of any nation-state, so there are no governments to apply to.

3 Do I need travel insurance to go to space?

You don't need insurance, but your 'vehicle' will need a heck of a lot. Before your trip, one country will accept responsibility for your spacecraft when it is logged with the United Nations Register of Objects Launched into Outer Space. They will be legally liable for any damage you or your spacecraft cause in space or by crashing to earth, so they will not allow you to get near your rocket without some solid insurance. To launch an unmanned satellite from the UK, for example, the British government currently requires insurance of around $600 million to cover preparation, launch, satellite life and eventual de-orbiting.

The concern of countries about their liability is illustrated by the case of Dennis Tito. Tito became known as the world's first space 'tourist' after paying $20 million to spend a week on the International Space Station in 2001. There are two habitable modules, one under the 'ownership/launching' of the US and one of Russia. The Americans were so concerned about the possibility of Tito getting hurt or causing damage in 'their' module that Tito was nearly restricted to the Russian 'territory'. An agreement was eventually reached to allow him to move between the two modules. (Presumably the cost of his flight included liability coverage.)

4 What happens if you commit a crime in space?

There is some precedence of criminal liability in outer space. In establishing the ISS it was decided humans would be subject to the jurisdiction of either Russia or the United States, depending on which 'module' they were in at the time. Eventually a waiver of liability was established. Although it is as yet untested, I suspect a crime committed by an astronaut would be similar to a diplomat accused of a crime in a foreign territory: the person in question would potentially be in trouble, though there would be some level of 'immunity' based on this waiver.

The social norms for good behaviour by astronauts and their small numbers make crime unlikely. What will complicate things is when more civilians go into space and journeys become far longer. If we take high seas law as a precedent, a crime committed in the neutral territory of outer space would be considered under the laws of the country that is the spaceship's 'launching state'. However, in the longer term some galactic norms and rules independent of earth's oversight seem inevitable.

5 Can I get married in space?

A wedding in space would most likely follow the precedent of the high seas: the captain of the 'vessel' could officiate, and the marriage itself would be subject to the laws of the country registered as the launching state of the spaceship. You would basically be getting married 'in' the country that 'owns' the spacecraft.

But my advice would be to have the wedding on land and a symbolic ceremony aboard your chosen craft. If you are a Russian cosmonaut, it might be written into your contract that you may not wed in space – a clause supposedly added after

Yuri Malenchenko married his wife (who was in Texas) in 2003, which garnered disapproval from the Russian authorities.

6 What legally happens to my corpse if I die in space?

Your death would probably be recorded according to the domestic laws of the country that launched your spacecraft. If your body were then expelled into space (which seems likely on a long trip, such as to Mars), I believe it would remain the legal responsibility of the state that launched the spacecraft. If your body later collided with another spacecraft, the launching country would be liable for the damage. If it were to collide with something else, there is a decent chance this would be known: we are quite good at tracking large pieces of debris we have put into space. However, so long as your body were either jettisoned into deep space or put into a 'decay' orbit to burn up in earth's atmosphere, rather than left in orbit, it is highly unlikely it would encounter another object.

7 What is the nationality if a baby is born in space?

This has yet to happen, but I suspect it would be a combination of the mother's nationality, the

father's nationality and the nationality of the spacecraft that the birthing mother is travelling on. Babies born on aeroplanes travelling internationally are subject to various combinations of these three. Where I suspect this will get more interesting and controversial is when we are in a long-term flight or colonization situation. What if the parents no longer feel allegiance to any country and want their child to be not only stateless but perhaps 'earthless'?

8 Do I need to file taxes in space?

Astronauts are not exempt from paying taxes and will be liable to file in whatever country they are domiciled in on earth. Apollo 13's Jack Swigert famously asked Mission Control how to apply for an extension on his income tax return. But where an astronaut is technically from is not always clear. Several NASA astronauts were born elsewhere but became US citizens to apply to the American-only space programme. Where a person pays taxes depends on the laws of the country they are born in and that of the country they have relocated to. However, being off-earth is not an excuse.

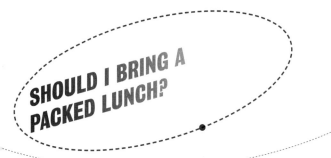

En route to the moon, our Spanish explorer Domingo Gonsales was visited by a number of 'Devills and wicked spirits' who conveniently supplied him with food for the journey:

They did so, readilly enough, and brought me very good Flesh, and Fish, of divers sorts well dressed, but that it was exceeding fresh, and without any manner of relish or salt...

For the rest of us, without such supernatural suppliers, we'll have to take what we need with us. Or produce it ourselves as we go along. The longer you stay in space the more important those food considerations are going to be. The early 'toothpaste food' space missions of the 1960s lasted for only a few hours or days – the length of a picnic or a camping trip. Now as we extend our trips to months and years, food needs to support not only the corporeal but also the mind. The psychological importance of a good meal, as you know from your own experience, is vital.

HOW TO MAKE A SANDWICH IN SPACE

Two days before the flight of Gemini 3 in 1965, astronaut John Young popped into Wolfie's Restaurant and Sandwich Shop at the Ramada Inn, Cocoa Beach and bought a corned beef sandwich. Along with testing the new two-man Gemini spacecraft, they were also tasting the latest in flavoured gelatinous

food cubes. On board he suddenly produced the sandwich from his spacesuit pocket, but as the mission transcript shows, its structural integrity didn't hold up to the rigours of spaceflight. The contraband sandwich is now forever preserved in the Gus Grissom museum in Indiana.*

Despite the NASA reprimand, a serious piece of knowledge was learned: crumbs in space could interfere with equipment and be a choking hazard. These days on board the ISS the 'space sandwich' conundrum is solved by using thin, crumb-free flour tortillas – the type you'd find in a burrito, and packed in oxygen-free bags so they'll last for a couple of years or so. The humble tortilla is also a good structural base to hold everything you want to eat together. Much more practical than a plate that relies on gravity to work and needs to be washed up.

COOKING IN SPACE

Space food has come a long way in sixty years. On board the ISS, meals are prepared on earth and then canned or bagged in single serving pouches like cat food to be reconstituted or reheated on the station. There isn't a standard kitchen – no open flames, running water or fridges or freezers. There are no waste food composting bins for fruit flies to gather round. The international crews all have their own food suppliers, but

---

* Address: 3333 State Rd, 600 East, Mitchell, IN. Phone Number: +1 812-849-4129.

wolfie's RESTAURANT & SANDWICH SHOPS, MIAMI BEACH

CONFIDENTIAL 45

CYI-2

1 50 49	C	Stand by.
1 51 06	P	If either one of these things leaks, we can just close up shop. (Food packages)
1 51 15	C	Okay. What's the Two-Bravo time?
1 51 18	CC	Roger. ΔV 90. GMTRC 16 52 25. GETRC 02 28 35. Roll left 55.
1 51 45	C	Roger. Two-Bravo: 90 ΔV. 16 52 25 GMTRC. Elapsed time of 02 28 25. Roll left 55.
1 51 55	CC	That's affirm.
1 52 26	C	What is it?
1 52 27	P	Corn beef sandwich.
1 52 28	C	Where did that come from?
1 52 30	P	I brought it with me. Let's see how it tastes. Smells, doesn't it?
1 52 41	C	Yes, it's breaking up. I'm going to stick it in my pocket.
1 52 43	P	Is it?
1 52 49	P	It was a thought, anyway.
1 52 51	C	Yep.
1 52 52	P	Not a very good one.
1 52 54	C	Pretty good, though, if it would just hold together.
1 53 13	P	Want some chicken leg?
1 53 15	C	No, you can handle that.
1 53 23	C	What was the time of that booster again? What elevation?

This corned beef sandwich, preserved in resin, memorializes the infamous sandwich incident during the Gemini III flight. John Young, co-pilot, snuck a corned beef sandwich from a favorite deli into space, by concealing it in his spacesuit. He pulled it out mid-flight and offered Virgil Grissom a bite.

SECOND STAGE: Clearing the Tower • **125**

foods are mixed, matched and shared as you would expect. Fresh fruits and vegetables can only arrive when the resupply ships arrive, and when they do crews gather around the hatch as it is opened, to take in the pungent citrus smells of the oranges. The smells of earth.

There are a multitude of practical considerations when preparing food for astronauts: nutritional needs, to address a plethora of health considerations; food consistency has to be just right, with not too much liquid that could escape; the food has to have passed stringent microbial checks, to make sure it doesn't go off or cause food poisoning. Salt levels are strictly controlled. Like air travel, space travel will affect your taste buds – one of the effects of microgravity is a redistribution of your body's fluids, resulting in your face becoming puffy and blocking your nasal passages and, like having a cold, affecting how you taste things.

GASTRONOMY DOMINE

Personal food preferences are also important. Each astronaut has a choice of 'bonus food'. An Italian company named Argotec is responsible for the European astronauts' bonus food, and Space Food Systems Laboratory (SFSL) for the Americans – they have about two hundred different items on their specialist menu as well as commercially bought foods for astronauts to choose from. Argotec have a 'ready to lunch' website where you can buy the same food for yourself. They have also developed an espresso machine for the ISS with coffee company Lavazza. Italian astronaut Samantha Cristoforetti became the first astronaut to have a decent cup of coffee in space.

Space food is increasingly 'cheffy', with tie-ins to the latest Michelin-starred gastronauts. The ISS has become the most expensive and exclusive *Restaurant at the End of the Universe*. Tim Peake worked closely with chef Heston Blumenthal, who is happiest

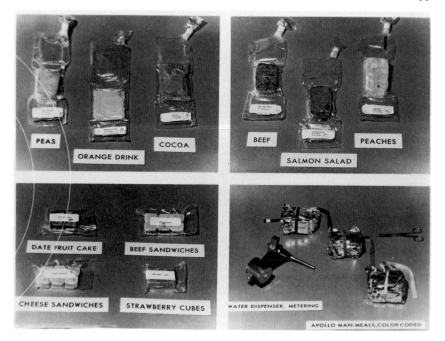

working at the more experimental end of food science. The crowning glory of Heston's work was the crumb-free canned bacon sandwich that was waiting for Tim when he arrived. The bar is constantly raised in the world of space gastronomy – chef Thorsten Schmidt worked with Danish astronaut Andreas Mogensen to create space-rock chocolates. Each one, like a fortune cookie, had a surprise handwritten note inside from a loved one.

LIVING OFF THE LAND

As we venture further afield, bringing all our food with us from earth will no longer be practical. Scientists have been thinking about this problem for a long time – greenhouses and hydroponic growing methods, genetically modified crop strains and experimenting with growing in various simulated alien environments. But there is much to learn – the first off-planet space lettuce was grown and eaten on board the ISS only in 2015. If you're familiar with the book or film *The Martian*, you'll recognize this theme – the astronaut castaway survives on potatoes painstakingly nurtured in the sterile Martian soil using his faeces as fertilizer. Astrobiologists at NASA Ames Research Center and the International Potato Centre in Peru have been working together on just this idea (without the faeces), trying to understand the minimum conditions that a potato needs to survive. Special environmental 'cubesats' have been built recreating the Martian environment, with soils used from the arid Peruvian deserts. Excitingly, you can see how the potatoes are doing LIVE on their website.

If we are truly going to leave the cradle we'll have to think beyond the packed lunch and learn how to feed ourselves by living off the land. In the meantime here's something to keep you going. The actual Space Rocks recipe, as flown to the ISS, for you to make at home. Kindly shared by Thorsten Schmidt.

Above: **NASA astronaut Peggy Whitson's tortilla cheeseburger**

SPACE ROCKS

Created by Thorsten Schmidt, 'The Nordic Alchemist',
for the ESA astronaut Andreas Mogensen on board the ISS.

Group A:
500 g dark chocolate (70% Friis-Holm
Danish chocolate)
400 g cream 38% (in this case high
temperature processed / long
pasteurized cream)
100 g espresso (Sigfreds coffee from
Denmark)
50 g of sorbitol

10 g rum
40 chocolate shells / spheres

For ending texture:
500 g of temperate chocolate
100 g crushed TUC Original biscuits
100 g popping candy

For visual colouring:
2-3 g carbon black
2-3 g natural white mineral

1 Mix all ingredients in Group A and temper to 72°C for
5 min.

2 Cool the chocolate mass slightly.

3 Stir the lightly chilled chocolate (28-30°C) with the
rum.

4 Fill the chocolate mass in the chocolate shells and
close the hole with a little tempered chocolate. Let
them cool for 30 minutes until the chocolate in the
balls has settled.

5 Roll the stuffed chocolate balls into tempered
chocolate, then crushed TUC biscuits and popping
candy. Make sure to cover chocolate balls completely
with the tempered chocolate so there are no holes.
Leave the chocolate ball to dry fully.

6 Now use a dry brush to colour the chocolate balls
with black and white mineral (food colour in powder
form), until they look like space rocks.

7 Space rocks are now ready to be served.

HEALTH AND SAFETY

'When Man steps into his rocket ship and leaves earth behind, he must be well equipped to survive in the hostile realm of outer space. To portray the complex problem of space medicine, we have designed a sort of common man. A man just like you and me. We will find out what will happen to him in a trip into space. In a way he's going to be our space guinea pig. That makes him a brand new biological species. I think we should call him *Homo sapiens extraterrestrialis.* Or, Space Man…'
Dr Heinz Hoffmann,
Walt Disney's *Man in Space*

RISK ASSESSMENT

Wherever you're planning to go in space, even if it's just the few hundred kilometres to low earth orbit, bad things can happen. If it's a cliché that there is nothing routine about going into space, it's a cliché for a reason. Many minor and not so minor incidents have taken place over the years, many of which you won't have heard about and have gone unreported, or have been hidden behind walls of political secrecy. Thanks to Ron Howard and Tom Hanks, you'll be familiar with the Apollo 13 incident – an explosion en route to the moon that became one of NASA's finest hours. Others not so: the Space Shuttle Columbia accident, caused by a piece of insulating foam coming loose on launch, striking and damaging the leading edge of the wing, which led to it disintegrating on re-entry. And Challenger breaking up soon after launch, caused by a malfunctioning frozen rubber 'O ring' on one of the solid rocket boosters.

Further back was the Apollo 1 fire that killed astronauts Roger Chaffee, Ed White and Gus Grissom, an accident preceded by the similarly grisly death of cosmonaut Valentin Bondarenko – one of the first group of cosmonauts – in 1961, only weeks before Yuri Gagarin's historic first flight. The cosmonauts would train in the sensory deprivation isolation room known as the Chamber of Silence. Bondarenko was coming to the end of his ten-day training stint living in the chamber and was cleaning the patch of skin where the ECG medical sensor had been attached with an alcohol cotton swab. Tossing the swab away, it landed on the small electric cooking hotplate, which immediately ignited the oxygenated surroundings, engulfing him in flames. The pressurized environment meant the two doors, like submarine hatches, couldn't be opened immediately. Several minutes later he was pulled out of the chamber, still conscious but his body almost completely burned. Only the soles of his feet remained unburned, where doctors tried to administer intravenous lines to give him painkilling treatment. He survived in the utmost agony for a few hours. His death was only reported in the 1980s. A crater on the moon is now named after him. In 1967, Vladimir Komarov was killed when his Soyuz spacecraft parachute failed to open, and four years later Soyuz 11's explosive decompression on re-entry killed the crew Georgy Dobrovolsky, Vladislav Volkov, and Viktor Patsayev. Recently Virgin Galactic's SpaceShipTwo crashed over the Mojave Desert, killing the co-pilot Michael Alsbury.

Bad things will continue to happen as we push further into space. But the fact that there hasn't been a single fatality on board any space station during the last forty years is a tribute to how far we've come.

STAYING HEALTHY – A NOTE FROM YOUR DOCTOR

If you're going on an extended adventure anywhere, it's always wise to seek medical advice. Luckily for you, I've consulted a doctor on your behalf. Here are some thoughts on the problems of staying healthy in space from medical doctor, cosmonaut and ISS Commander Dr Oleg Kotov.

Dr Oleg Kotov, MD
Cosmonaut and ISS Commander
June 2017, Moscow, Russia

What is the most difficult thing in space from the point of medicine? It is not possible to have a fully staffed polyclinic* with all the necessary equipment on the spacecraft. How to solve this problem, especially for flights beyond the Earth's orbit to the moon and Mars?

Firstly – the selection and the medical screening of astronauts should be more thorough.

Secondly – it is necessary to determine the list of the most probable illnesses and diseases that can arise in flight and to provide the necessary medicines and equipment for their treatment. And hope that no one will become sick with an illness that is not on the list.

Thirdly – we need to learn to conduct surgical operations in weightlessness (which we have not yet begun to do).

And in the fourth place, the most important thing is to prepare an astronaut doctor and medically train one more member of the crew, a paramedic for all possible medical scenarios. Of course there are opportunities for telemedicine and consultations with doctors on Earth – but as the distance from our planet will keep increasing, this thread will become thinner and thinner.

Another pressing problem is weightlessness. Until artificial gravity is created in one form or another, we will constantly struggle with the adverse effects of weightlessness on the human body. Special suits, many hours of exercise, medication and other tricks help overcome the effects of weightlessness, but not completely. This problem is yet to be solved and must be solved by science.

Radiation – in the conditions of long flights to the moon and Mars outside the Earth's radiation belts, with the subsequent long stays on their surface, the development of a means of protection from the effects of radiation on humans and technical systems is one of the most important tasks.

Not entirely medical, but a problem close to it in spirit: the psychological compatibility of the crew, and social and psychological support**. When immediate return to earth becomes impossible, the crew will need to have as much information as they need to support decision making. These decisions will affect the success of the entire mission and even the lives of people on the ship. The role of the crew commander will be especially important in those circumstances. The compatibility of the team members and their unity should be nurtured long before the start.

The problem faced by crews in long duration flights in the earth's orbit is of microbial and fungal contamination of hermetic volumes. Despite the observance of all quarantine procedures and special measures, a significant number of species of microbes and fungi become passengers on spacecrafts and stations. They settle on panels, cable harnesses, equipment, in garbage bags, dirty linen bags, etc. All existing methods of combating this phenomenon have shown their inefficiency. We need to develop a method for disinfection of space objects, taking into account that within these objects people are working and living.

---

* This is a Russian term for a clinic, where outpatients are treated. It is usually a multi-storey building with general and very specialized medical personnel. I haven't come across this term being used in UK.

** Russian crew also receive 'information support', I guess the closest term would be the 'psychological support through information'. This could be updates in specified areas and news close to the interests of the crew member.

CAN I BUY A TICKET TO SPACE?

Perhaps the idea of waking from a 600-year cryosleep searching for a new planet in some far-off solar system isn't for you. A three-year round trip to Mars might be too much of a strain. Perhaps a journey of just a few days, or even a few minutes, is all you need to scratch your itchy feet. The promise of a short package holiday away from earth is something that we've often been told is imminent, but we're living in a time where this could actually be a reality – where a trip into space isn't just for the professionals, but for you. And by *you*, I mean someone a lot richer than you.

In 1968, Apollo 8 took humans to the moon for the first time. Just as significantly, that year saw the release of Stanley Kubrick's visionary interpretation of Arthur C. Clarke's *2001: A Space Odyssey*, a film which inspired and influenced a generation of astronauts and space travel visionaries. It depicts a future of routine trips to the moon, with pristine turbaned stewardesses, and exciting trays of space food. The film merged the reality of the moon as a destination with a new age of aviation and travel glamour that had landed on earth – the brand-new spacious Boeing 747 that would bring jet-set travel to the people was about to come into service and shrink the world. The message was clear – flying to the moon would soon be as accessible as flying across the Atlantic.

In that same year, capitalizing on this new optimism, the American airline Pan Am, who we see in the film, started its 'First Moon Flights Club' back on earth. Pan Am was at the forefront of the new dawn of aviation expansion, with a 'meatball' logo just as iconic as NASA's own. A quick call to your travel agent would get you on the Moon Flights Club list, which over the years grew to 100,000 members. Of course there were no firm plans. There was no Orion 3 spaceplane, the spacecraft we see in the film. Instead you received a card and a letter of intent that this would one day happen – despite fares that 'may be out of this world'. This was one of the first loyalty cards. Pan Am knew what people wanted, and knew how to tap into that dream.

Know All Ye by These Presents that

Jeffrey Gates

has become a certified member of Pan Am's

"FIRST MOON FLIGHTS" CLUB

1043 _James Montgomery_

Number   Vice President, Sales

Above: **The two 'meatball' logos that changed the twentieth century**

ASTRONAUT WANTED: NO EXPERIENCE NECESSARY

In the 1980s, the reusable Space Shuttle ushered in a world of new opportunities. With more seats on board than any other spacecraft before, it paved the way for access to space for a more diverse group of people, including, for the first time, 'citizen observers' – most famously Christa McAuliffe, the New Hampshire schoolteacher who flew on Challenger STS-51-L in 1986, who tragically died along with her six other crewmates.

As the political landscape changed in the Soviet Union during the years that followed, the cash-strapped Russian space programme announced that their Mir Space Station was opening its hatch for business and accepting paying guests. In 1990 Toyohiro Akiyama, a Japanese journalist, became the first commercially paying civilian in space. According to Russian space historian Anatoly Zak, a Japanese broadcasting crew from TBS (Tokyo Broadcasting System) had been covering a launch at Baikonur and had joked to their host about the possibility of hitching a ride. They were more than surprised when the answer they got back was a 'Da'. Despite apparently chain-smoking four packets of cigarettes a day, Akiyama was selected out of 163 other applicants, along with a camerawoman called Ryoko Kikuchi who later had to pull out due to an illness. Giving up the booze and cigarettes, Akiyama underwent the formal training at Star City in Moscow, including the daunting task of having to learn Russian. The week-long trip on the Mir Space Station cost $11 million, and was sponsored by the TBS network. Despite suffering from severe space sickness Akiyama performed his duties as a journalist, photographing the earth and broadcasting his overview effect experiences back to his public on earth. He described the view of the earth from space as 'like listening to an orchestra of colours'.

Britain also bought a visit to Mir through a consortium of commercial investors, which became known as Project Juno, named after the Roman goddess whose temple was guarded by geese. Chemist Helen Sharman was selected for the trip after hearing the job advert on the radio – 'Astronaut wanted. No experience necessary' – and hastily scribbling down the phone number on a petrol receipt while sitting in a traffic jam. Thirteen thousand people applied for this rarest of rare opportunities. Helen became Britain's first astronaut and spent eight days aboard the station in May 1991, doing a variety of scientific experiments and outreach work.

At the end of the 1990s, the creaking Mir Space Station was being increasingly exploited as a commercial platform. The American TV network NBC had even greenlit a reality show called *Destination Mir*, in which contestants would train as astronauts, with the winner launching into space. The idea was set up by MirCorp, the maverick private space company who had leased the station from the Russians. The show never made it off the ground and Mir was finally decommissioned in 2001.

Above: **First British astronaut, Dr Helen Sharman**

С. КРИКАЛЕВ **Х. ШАРМАН** **JUNO** **А. АРЛЕБАРСКИЙ**

BUYING YOUR OWN TICKET

It was Dennis Tito, a millionaire and former NASA Jet Propulsion Laboratory engineer, who would eventually claim the honour of being the first self-funded astronaut. His trip was made possible through Space Adventures, the first and only space travel agent, who right now are waiting to take your money and organize your trip. Tito bought a seat on board a Soyuz to the brand-new International Space Station. This went ahead despite the protestation of NASA who weren't convinced that Tito was ready and thought he could be a safety issue. His trip marked the beginning of new possibilities for access to space for a very select few – a small group of committed super-rich individuals, who had the opportunity, passion and financial resources to make their dreams come true.

Even being a multimillionaire doesn't guarantee you a ride in the coveted right-hand seat of a Soyuz. Successful training has to be completed, the Russian language has to be learned, and exams and medicals passed. Sarah Brightman, the singer, paid $52 million for her ten-day mission to the ISS, which was to include singing a specially composed song in space. But despite all her training, she was forced to pull out for personal reasons.

The British/American video-game pioneer and entrepreneur Richard Garriott lifts up his shirt and shows me a huge scar across his belly. We're in the Duke of Yorke pub in Holborn and I'm presenting him with his replacement silver British astronaut pin. Richard ran into medical problems during training for his self-funded flight. Russian flight doctors had identified a problem with his liver. It was so minor that he may well have lived his life never knowing about it, but it was enough of a risk for the doctors to ground him – despite having already paid millions of non-refundable dollars. On hearing the news, Richard took matters into his own hands, and his own doctor in America went ahead and surgically removed a sizeable chunk of the offending organ. When he came around from the anaesthetic, his doctor informed him that the operation was a success and his liver was in good shape. He pointed out that he'd also noticed something odd about his gall bladder while he was in there, so had taken the liberty of whipping that out too. Richard asked the doctor if he'd checked that this unscheduled medical procedure had been approved by the Russian flight doctor back in Star City. Do you need a gall bladder to go to space? It hadn't even occurred to the doctor to ask. Panic ensued, phone calls were made, but eventually Richard got signed off to fly.

THE ORIGINAL SEVEN SELF-FUNDED ASTRONAUTS

Name	How they paid for their ticket	Flight duration	Date	Return ticket*
1. Dennis Tito (USA)	Investment Management	8 days	28 April–6 May 2001	$20 million
2. Mark Shuttleworth (SA/UK)	Internet security certificates	11 days	25 April–5 May 2002	$20 million
3. Gregory Olsen (USA)	Optoelectronic sensors	11 days	1–11 October 2005	$20 million
4. Anousheh Ansari (US/Iran)	Software	12 days	18–29 September 2006	$20 million
5. Charles Simonyi (US/Hungary)	Microsoft Office	15 days 14 days	7–21 April 2007 26 March–8 April 2009	$20 million $35 million
6. Richard Garriott (US/UK)	Video games	12 days	12–23 October 2008	$30 million
7. Guy Laliberté (Canada)	Founder of Cirque du Soleil	11 days	30 September–11 October 2009	$40 million

* All-in. Includes travel, accommodation, transfers and all meals. Non-refundable. Don't forget to read the small print.

INCENTIVES – THE X PRIZE

Until recently, national government controlled space agencies were the only ones holding the keys to space. For the space frontier to be truly opened to all it needed an incentive, something tangible for the dreamers and entrepreneurs to get behind. While political conflict provided the impetus to get us into space in the 1960s and 1970s, now it was about the money.

Incentive prizes of the past, which offered cash, like John Harrison's Longitude Prize, or the Orteig Prize won by Charles Lindbergh for flying non-stop from New York to Paris, have made a return to help drive exploration and innovation. It was reading Lindbergh's book *The Spirit of St Louis* that inspired Peter Diamandis to create the X Prize in 1995, to 'build and launch a spacecraft capable of carrying three people to 100 kilometres above the Earth's surface, twice within two weeks'. Domingo Gonsales and his flock of geese (even though he had built his own spacecraft) would have sadly been ineligible because his 'flight of fancy' was a one-off.

A whole world of different solutions to Diamandis's challenge appeared: space planes, 'rockoons',* and vertical take-off and landing rockets. It was aircraft designer and entrepreneur Burt Rutan's SpaceShipOne, with its revolutionary shuttlecock folding tail section used to slow it down on re-entry, that would claim the prize. On 21 June 2004, and again on 29 September, test pilot Mike Melvill flew this pioneering new spacecraft into history. At 47,000 feet, SpaceShipOne was released from the carrier aircraft, and its rocket engines were lit sending Melvill straight up at over 2000 mph. He said, 'It was like being hit with a sledge hammer in the back.' Melvill, and later Brian Binnie, had crossed the Kármán Line, making them the only two people so far to receive FAA astronaut wings as commercial pilots.

SOMETHING FOR EVERY BUDGET

MOON
COST: $$$$$$

SPACEX'S DRAGON 2

The new Chief Designer, SpaceX's Elon Musk, has just announced that two self-funding space adventurers have signed up and paid the deposit for a week-long trip around the moon on board his Dragon 2 spacecraft, currently slated for the end of 2018, although I suspect this date will change. Those two people are _____ and _____, and between them they've forked out an eye-watering $____ million.** This is SpaceX's first foray into the commercial 'space tourist' business. If it happens, SpaceX could pip NASA's Exploration Mission-2, a similar route around the moon on board the new Orion spacecraft launched on NASA's forthcoming SLS rocket, due to be launched around 2022 – again, subject to delays. Both these projects represent a huge step forward in human space flight: the first time we will have gone back to the moon for half a century.

But don't worry if you don't have a spare $____ million. There are other options on the horizon…

---

* 'Rocoons' are rockets carried by balloons that are launched from high in the atmosphere.

** Please fill in the blanks.

SUBORBITAL
COST: $$$$

The two frontrunners in the race to suborbital space are Virgin Galactic and Blue Origin. Both systems – the reusable spaceplane of Virgin and the ballistic capsule design of Blue Origin – along with SpaceX are the most visible part of what's dubbed as the NewSpace culture. A drive for space access led by entrepreneurship.

VIRGIN GALACTIC –
WHITE KNIGHT/SPACESHIPTWO

Since the X Prize was won in 2004, Richard Branson has been trying to get his own Virgin-branded reusable tourist spacecraft up and running. But as with any new rocket technology, this has been beset with technical, bureaucratic and political problems.

The idea is for a suborbital flight for tourists – a lot simpler than the speeds and complexities needed for full orbital spaceflight. This will take paying customers

on a huge parabolic ride, crossing the 100 km Kármán Line, before landing on a runway back on earth. The apex of the parabola, just like on my roller coaster, will give the astronauts their few precious minutes of weightlessness.

A jet-powered aircraft (WhiteKnightTwo) carries the rocket-powered spacecraft high into the atmosphere to 50,000 feet, almost the limits of a jet engine, before releasing it. The spacecraft then fires its rocket motor, and heads straight up to cross the all-important line before gliding back to earth, ready for you to show off to all your friends.

On the Virgin Galactic company's website you will be asked: 'What are your motivations to go into space?' Hopefully this book will help you come up with a pithy answer. Tucked away at the bottom of the webpage is: 'By ticking this box you are confirming you are over the age of 18 and understand a spaceflight requires an upfront deposit of US$250,000.00.'

BLUE ORIGIN –
NEW SHEPARD – 'Gradatim Ferociter'*
Jeff Bezos, the CEO of Amazon, the man who delivers your stuff in brown cardboard, has his own solution to deliver *you* into space.

In 1946 the British Interplanetary Society came up with a project called 'Megaroc' – a two-stage, V-2-inspired, ballistic suborbital craft that would propel a crewed capsule over 300 km. It's an idea that never made it past the drawing board, but it's an almost identical concept to that of Blue Origin. Here's the plan: a vertical take-off vertical landing (VTVL) rocket booster with a big feather painted on the side called New Shepard (after Alan Shepard) will launch a crew capsule above the 100 km Kármán Line. Most importantly, the capsule you travel in will have room enough for you and five friends to do your somersaults, with big 28.6 x 42.7-inch windows where you can bear witness to a universe of overwhelming indifference. You will then come back down to the ground suspended from three parachutes.

The company's coat of arms features two awestruck tortoises (reminiscent of those who went to the moon) reaching for the sky holding a shield. Above the door of the crew capsule, the theme continues: a tortoise picture is stamped to mark every successful test flight.

Price: TBC

* 'Step by step, ferociously.'

BALLOON
– THE ORBITAL PERSPECTIVE
COST: $$$

These days it's de rigueur to send unusual things like meat pies to the 'edge of space', tethered to helium-filled weather balloons. On *Bang Goes the Theory*, we even sent up a small 3D-printed doll version of me, which ended up landing in a Travis Perkins builders' yard somewhere near Cambridge. The closest I will ever come to spaceflight.

The 'edge of space' is a bit of a cheat because it isn't really the edge of space at all – 40 km or so is the limit of a weather balloon. The pronounced curvature of the horizon you see in many of these photographs is as much an effect of the camera lens as altitude. In fact the term 'edge of space' has become a catch-all for anything above 60,000 feet (18 km) or so, getting on for twice the cruising height of a commercial airliner. Here the air is so rarefied most jet engines are no longer of any use – the only way to access this realm, other than with a rocket, is with a gas-filled balloon.

Plans are afoot to send humans up here, and not for the first time. Project Excelsior and Project Manhigh were US Air Force high-altitude balloon programmes from the late 1950s and early 1960s that took Colonel Joe Kittinger to 102,000 feet (31 km) before he jumped out, breaking the altitude freefall record. This longstanding record was beaten in 2012 by Felix Baumgartner and then again by Google executive Alan Eustace in 2014.

Companies such as Zero 2 Infinity are now developing edge-of-space balloon rides

Above: **Meat and potato pie in the sky**

offering much cheaper access for would-be astronauts looking for the 'orbital perspective'. Ron Garan is a veteran Space Shuttle pilot who is the chief pilot and public face of World View, a company that sends paying customers into the stratosphere. Their balloons can be used as scientific platforms and 'stratolite' delivery systems (satellite, but in the stratosphere). But most interesting is the Voyager capsule – a 'comfortable, stylishly-appointed spacecraft' that will get you to heights very few have visited. For two hours you'll be able to experience your own overview effect, gently suspended at the very top of the atmosphere, at the limits of where a balloon can reach. It will also be the first spacecraft-rated vehicle to have a dedicated fully-stocked cocktail bar on board.

Price: $75,000. Coming soon.

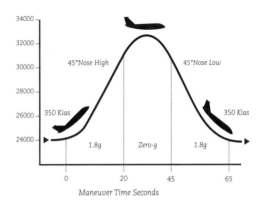

Altitude, Feet

'VOMIT COMETS' AND FOXBATS
COST: $$

If you suffer from motion sickness, then make sure you take an appropriate pill and something to throw up in. The 'vomit comet' is a commercial aircraft that flies in a series of parabolic arcs. At the apex of each arc you become weightless, until the aircraft goes over the top and begins to dive again where you will feel 'hyperschwerkraft'* – about 1.8-g – the same way you experience weightlessness on a roller coaster, or your stomach goes tingly when you drive fast over a hill, but obviously a much bigger 'hill' giving you a period of much longer weightlessness. For about twenty seconds per arc, you're experiencing what it's like in orbit – one endless arc. What you'll need is something roomy like an Airbus 300 or a Boeing 727, or even an old Soviet Ilyushin il-76, which you could pick up pretty cheaply these days. Get them to take the seats out so you've got lots of room to float about in. Luckily for you, there are now various companies like Air Zero G in France who have arranged all this for you. Then find a brave pilot with a strong stomach. Zero-g flights have long been a useful analogue for astronauts to train with, as well as a research platform for experiments that require microgravity.

For €20,000 you can get a ride into the stratosphere on board a Russian Mig 'Foxbat' fighter jet, a service that several private Russian companies provide. Four hundred kilometres from Moscow is the Nizhny Novgorod air base. From here your friendly Russian pilot will take you to around 18 km. Up there at the top of the atmosphere, through your all-glass canopy, the sky will darken to almost black and the curve of the horizon will be seen – a tantalizing hint at what's in store if only your pilot could take you higher.

---

* For our German readers.

SPACEPORTS
– WHERE DREAMS ARE LAUNCHED FROM
COST: $

Pancho Barnes's Happy Bottom Riding Club hasn't been serving up its legendary free steak dinners since a fire razed it to the ground in 1953. Pancho was the barnstorming female aviator who bought a ranch near what was the Muroc Air Force Base before it grew into Edwards Air Force Base, where the Space Shuttle sometimes landed. She set up the restaurant and bar hangout beloved of the test pilots and made famous in Tom Wolfe's *The Right Stuff*. The first free steak dinner was given to Chuck Yeager in 1947 after he finally went through the Mach 1 barrier in the Bell X-1. The steak dinner prize was subsequently offered to every pilot who did the same. The ruins of Pancho's club are slowly being eaten away by the desert. A few foundations, the odd wall and the old swimming pool are up there in Mojave, just off the road that leads to Edwards, a few hours drive north of Los Angeles.

Out here is another important space hub – The Mojave Air and Space Port. For the last two decades this facility has been the beating heart of the NewSpace movement, the home for the new private rocket-ship builders, who risk it all to try to leave the planet. It's a wild west, tumbleweed kind of place. As you drive in, you're greeted by one of the monuments to the early days. The 63-foot tall Roton Rocket stands like an obelisk, once the 'future' of reusable spacecraft – designed with rotorblades at its tip so it could auto-gyrate back to earth. A good idea, which like many good ideas, was hobbled because the world just wasn't ready for it. Men like Jeff Greason – one of the early pioneers – packed up his life and moved out here to pursue this dream. Having worked on the Roton, he moved on to co-found the XCOR Aerospace company, which, like Branson and Bezos, is trying to help you get into space as cheaply as possible.

The Mojave Air and Space Port even has its own restaurant, The Voyager Restaurant Diner, which proudly advertises, 'Aviation spoken here'.* Not quite as legendary as Pancho's, but you might bump into one of the NewSpace pioneers, like Dave Masten whose Xombie rocket I watched being tested here a few years back – a stripped down, alcohol-fuelled prototype balancing on a column of hypersonic gas, that landed back on its feet long before Elon Musk's Falcon 9 stunned the world doing the same. Maybe if you cross the Kármán Line

---

* Open Monday to Friday 7am to 3pm, Saturday 8am to 3pm and Sunday 8am to 2pm. Tel: +1 (661) 824 2048.

the Voyager diner might offer you a free steak dinner on the house. If they won't, I will.

As the finishing touches are still being finalized on the new generation of private space vehicles, it is worth thinking about where you are going to fly from. What should a modern spaceport look like? Grand, exciting, state-of-the-art. Something to rival the world's great airports. Something you'd commission the great Norman Foster and Partners to design, as your first-choice architect. The good news is that planet earth's first purpose-built commercial spaceport is ready. Looking like a horseshoe crab, or a crashed flying saucer clinging to the New Mexico desert near the town of Truth or Consequences,* Spaceport America is a grand, exciting, state-of-the-art building designed by Norman Foster and Partners. This is the shiny new home for Virgin Galactic, as well as a test facility for SpaceX and others. They have their own fire department, twenty-four-hour security, a medical centre and huge hangars to park your spaceship. Being the *only* spaceport limits your destination choice. But that will hopefully change one day too.

If you're in a hurry, you can visit it right now and do the official tour. There won't be any flights leaving yet, but you can always turn up early and be first in line.

Above: **Spaceport America, New Mexico**

---

* The NBC radio show Truth or Consequences offered to broadcast its tenth anniversary show from the first town that would legally change its name to Truth or Consequences. Hot Springs won.

TRAVEL GUIDE: THE ASTRONAUT

Name.
Alfred M. 'Al' Worden
Occupation.
Test pilot, engineer, astronaut.
Claim to fame.
Apollo 15 Command Module pilot. Remained in lunar orbit conducting scientific work while Jim Irwin and Commander David Scott were on the moon's surface.

BEING AN ASTRONAUT
Did you see yourself primarily as an explorer, an ambassador, a scientist, a test pilot or an engineer?
I considered myself probably more of a scientist than anything else. The Mercury 7 astronauts were picked because they were test pilots – guys who would stand the best chance of surviving in space, not knowing what they're gonna find when they got up there. And that was very successful. By the time they got to my group, academic background was almost as important as your flying background – I was teaching at the Test Pilot school at Edwards Air Force Base, but I already had three Master's degrees from the University of Michigan, and that made me the perfect candidate for the program. As pilot, I did all the piloting going out and back. That's a skill that you learn, it's not an intellectual pursuit, it's like learning to drive a car, like learning to fly an airplane – you learn the systems, you learn what to do if something goes wrong, but it's a skill – it does not require a lot of knowledge or intelligence. The science part did require that, like the remote sensing of the lunar surface and the photography that we did on the flight. It required interpretation of what we saw.

Did you discover a love for lunar geology?
No question about it. When we made our flight, there was still a question of what made the features on the moon: was it meteor impact, or volcanic activity? Of course it's a little of both.

As a rookie, how was your relationship with the Project Mercury guys?
Hah! I will never forget walking into the astronaut office the first day I got there and being treated like I was a janitor. The older guys, those who had flown, kinda looked down their nose at the new guys coming in, because they hadn't flown. You really don't count for anything until you've made a flight. There were only 25 of us in the office when I got there. I was number – I dunno, I forget my number now, but there were a couple of Mercury guys left, there were quite a few of the Gemini guys left. We were the go-fers for everybody above us – we would go-fer the coffee and sweep the rooms. And then when I got assigned on the backup crew for Apollo 12 everything changed. We were part of the group then.

Talking about Apollo 12, would you be able to point to the SCE to AUX switch?
Dallas, my god man! I'm not sure I could do that anymore.

I think it was just behind Alan Bean's head.
Yeah. I think it's my 'age progression'. So I forget all those things.

I think you can get it on a T-shirt now. It's become a bit of a space geek thing.
I was out at the launch pad. I was the close-up crew guy, so I had to put all their straps on and connect the oxygen and radios. I was in the viewing area when Apollo 12 launched and they got hit by lightning.

Pete Conrad's hand would have been on the abort handle wouldn't it?
Yes indeed.

When you applied for astronaut selection, was it very similar to the Mercury guys? Did you have to do the Lovelace medicals and things like the Rorschach test?
We did the Rorschach, we did this, we did that. And I can remember distinctly asking the Air Force colonel who was in charge of all that, could they actually keep somebody out of the program based on psychiatric tests? And his response was: no we can't. All our reports go to the board, but we do not have the authority to say yes or no. However, if we declared somebody clinically insane, then that was as far as they would go. We had some guys that were on the borderline that actually got into the program.

What did you see in the Rorschach? Any squashed Zebras?
I saw a lot of butterflies. <Laughs> There's another test where they show you pictures – like a guy lying in bed in a jail cell and a bridge over a chasm going towards a castle. The last one was a blank white sheet and they asked Pete Conrad what he saw in it and I think his comment was, 'Just a lot of snow.' It was crazy kinda stuff.

Was it all very competitive?
Pete was pretty colourful. I don't think they selected Neil Armstrong because they thought he should be the first man on the moon, I think they selected Neil because he was in the right rotation. All of us in those days never, ever thought Neil would make a landing. It was always assumed that something goes wrong on a first flight like that. We never really gave it any credibility that Neil

would be the first one. We all thought that Pete Conrad would be the first.

WHY GO INTO SPACE?

Apollo was so much more than a political exercise, wasn't it?
Dallas, let me ask you a question. Philosophically. What do you think the goal of the space program is?

Now? Or then?
From the very beginning till now and way into the future. What is the ultimate goal of the space program?

Several things – an emotional pull towards the idea of 'the frontier'. To try and make sense of the universe…
Okay. Let me ask you the next question. What do you believe is the prime imperative of every living creature on earth? Everything – grass, trees. Animals. Humans. What is the prime imperative of all those living things?

Well, to survive. And reproduce.
It's survival! Correct. And I happen to believe that we have a genetic tick in our brain that says the day will come when we can't live here anymore, we gotta go somewhere else to live, so in my mind the ultimate goal of the space program is to find someplace else where we can go live.

And where did Apollo fit into that?
It's the first step. Forget about going to the moon. The first step is developing the technology to do that. The next step is to go further, the third step is to go further, the fourth step is to go further, and along the way the most important thing of all that's got to be developed is a propulsion system that will get us way out there, like Star Trek. We will not find another planet that we can live on until we get propulsion capability that will get us there in a reasonable amount of time. But you see, to me, the whole purpose of the space program, genetically driven if you will, is to give us the capability to take the human species somewhere else.

It's a recurring theme I keep bumping into – that notion that we need to be out there.
Well I think there's no question about it. And we've taken the first step in going to the moon. We couch these things in terms like 'exploration' and finding out what's in the solar system, and going to other planets and that kind of thing, so that gets people interested in what we're doing on a short-term basis. But in my mind, the long-term basis, and we're maybe even talking about a million years, I don't know, is to eventually give us the capability to go to

another solar system, where there's a planet that we can survive on.

LIVING IN SPACE

Jim Irwin and Dave Scott leave for the lunar surface, and then you're suddenly in this extraordinary position of being 'the world's most isolated person', as it's often described. Alone in lunar orbit.
You know, that's a bullshit record.

<Laughs>
But somehow that's one that gets picked up. It's kind of an interesting one. The one record I've got that nobody will ever beat is that I did the first deep space EVA.

Tell me about seeing the earth from the moon.
I saw the earthrise 75 times while I was there. I remember going to the window every time I went around the back side of the moon to watch the earthrise, because it's just unbelievable. Oh my god, that's 250,000 miles away, and I live there. Holy shit! And I can cover the Earth with my thumb. People ask, *'What was it like being up there by yourself for three days?'*, and *'Weren't you lonely?'* And I've gotta say, no. I wasn't lonely at all. I loved it. As a matter of fact, I had room to move around without those two guys there. I'd been with them four and a half days. I was so goddamn glad to

get rid of them. I thoroughly enjoyed the time by myself. And it allowed me to do a lot of things science-wise – I did a lot of photography of the lunar surface, a lot of remote sensing. I took pictures of things in the solar system that we'd only talked about, that we'd never proven.

Astronauts talk about the overview effect – this overwhelming sense of scale and connectedness with the cosmos.
I'm not sure I ever got that feeling. I used to look at the earthscape, and think oh I can see Florida, I can see the Gulf of Mexico, I know where Houston is, and then track down to the centre of my house. The Earth is so goddamn far away that it's just another object in the solar system. Even though you say to yourself, 'That's home.' Crazy.

How black is space?
Dallas, it's the blackest you've ever seen. I was in a part of the orbit round the moon where I was shadowed from both the Earth and the sun. Absolute complete blackness except for the universe out there. I used to stare at the Universe absolutely amazed because I could not pick out an individual star, there were so many of them. It was a wash of light that was continuous, and only a very few of the brightest stars could you see in that

background from all those stars that are out there. And if you start looking at what that all means – it tells you that we really don't understand the Universe around us very well. We live in the Milky Way galaxy. There are 400 billion stars in the Milky Way. And that's all you can see from the lunar orbit, the Milky Way. But you look beyond that and astronomers will tell you that there are a couple of hundred billion more galaxies like the Milky Way out there. The numbers get so huge that I can't comprehend them anymore. I could see the horizon of the moon, the mountain peaks, everything – not because you're in sunlight, but because of the background starlight they obscure. There were so many stars. A blank white light. Truly amazing.

People are drawn to astronauts because you provide a human link to something much bigger. Are you comfortable with that responsibility?
To be honest with you, I think it's great. Because I can say anything I want and who is going to dispute it? I mean unless you've been there you don't. There's no way you can question what I saw.

<Laughs>

NAVIGATING
They gave you a sextant as well as the primitive Apollo Guidance Computer. Did it makes you feel like a proper explorer using stars to steer by?
Well, the sextant we had was pretty much like sextants that have been used for a couple of hundred years, except ours was connected to the computer to calculate the angles; we didn't have to do that manually. You know, I navigated all the way back home on my own, I did not use Mission Control. I used the sextant to calculate where we were and derived the course corrections from that, and I got us back as well as Mission Control would have. In fact, we landed closer to our landing point than Mission Control had us. It was an interesting exercise.

Was that due to the technology or your skills as a navigator?
I don't think it was my skill. But there was a thought in my head: 'If we're gonna go to Mars, it's gonna be a year and a half,' and I got to thinking in a year and a half a lot of things can happen to electronics. So we've got to think of a way of getting back in the event of not being able to talk to Mission Control. So I convinced Mission Control and the guys there that we needed to give this a try. Mission Control had us accurately tracked,

but I decided that I needed to prove that an onboard crew, given the right set of conditions, could still get back home, and we did. We did it very successfully.

If somebody needs a ride to the moon, they should get in touch. You'd be the perfect lunar Uber driver.
Yeah! Ha ha. The perfect Uber. You know, when we go back to the moon, they will have a GPS system around the moon. So it will make it much easier to navigate back and forth.

If a moon tourist asked you, where would you want to take them?
I think you'd wanna take them to the South Pole.

Shackleton crater?
Yeah. The sense is there's water at the South Pole, so that would be an interesting place to go to. If you're just gonna go up there and camp for a while, one place is as good as another.

THE FUTURE
What do you think of Elon Musk? He seems to be like Wernher von Braun, both as a pragmatist and a visionary in the way he talks freely about big, lofty ambitious goals.
I think Elon Musk is a lot like Wernher in that respect. He is designing things that work – his returnable first stage rocket is unbelievable. I think

he's an amazing guy, he's very charismatic, he was able to raise money even though he didn't have anything to show for it. I'm a little concerned that there's this talk now how he's gonna send a couple of guys around the moon in the next few years, and there will not be a pilot on board.

Get in touch with him and say, I know the way. I've been there.
<Laughs> Damn right I'll take them! Well, if anything should go wrong, what are they going to do? We learned on our flight that robotic systems and artificial intelligence are great, but when something happens that they're not programmed for, then you've gotta problem. And we had those kinds of things happen on our flight, where a robot would never have figured out the problem.

Would you go back?
Goddamn right! Why not? I've had lots of people ask me, 'What kind of people should we send to Mars?', and I point to myself and say 'Me!' In the first place, I've lived a very good life, I'm an old guy now (although my doctor tells me I'm twenty years younger than my age). And as an old guy, I can sit and watch TV all day.

Yeah, there you go!
I don't care. <Laughs>

What do you think of the new NASA Orion spacecraft that's going to go around the moon? You can do a deep space EVA from the Orion like you did on the Apollo Command Module.
Yeah I know. My honest answer, it's a piece of shit.

Really?
It's just a repeat of Apollo. I think there's a huge, huge, deficit of creativity in NASA today. Now, the new SLS rocket might be okay, but I have a big problem with Orion. I took a group up to NASA headquarters trying to convince them of another shape, a re-entry shape that had all the capability in the world, but they had already started on Orion, they didn't want to talk about it. So they're stuck with old technology. Why is it that the Dragon works like Apollo, why does the Boeing Starliner look like Apollo? Why do all these machines look like the old Apollo? Because it's the most simple, minimalist spacecraft you can build. None of those vehicles are designed to go as far as Mars, they're all designed to work in earth's orbit or maybe go to the moon. That's it.

Back in 1638...
Holy shit! I wasn't alive then...

... Domingo Gonsales was pulled to the moon by a flock of geese in an early science fiction story.

I introduced him to my parents, by name: Tiny and Helen. And about three years later he was doing a series of talks through the country and he stopped at my hometown and gave a talk at the college there, Jackson Community College. My folks found out he was coming in town so they went to the talk. They were sitting about halfway back, he saw them there, and he called them out by name: 'Hey Tiny, hey Helen. Nice to see you.' And I thought, man, there's one slick politician.

He was a real showman, wasn't he?
Oh he was. Very charismatic. Very convincing and I think he made the whole thing possible.

There are great similarities between how we dreamt we would go to the moon and how we actually did it.
Well you know what, there's even a better one. Jules Verne wrote about three guys that went to the moon in a capsule, just like we did.

And they left from Florida!
Cape Kennedy or Canaveral was his launch site. He had that part of it all figured out, and he didn't know anything about rockets. Do you think that he was right about so much of what he wrote because he knew enough about it even back then? Or did somebody read his book and say, that's what we've got to do? I've never really resolved that question. But

he was way, way ahead of us. I've often wondered what it would be like to get fired out of a cannon at 25,000 mph. Would there be anything left of you? I think you might be a puddle.

I think a lot of early rocket design solutions came as a direct result of trying to solve Jules Verne's fundamental projectile design flaw.
Robert Goddard was part of that, Tsiolkovsky was part of it, Korolev was a big, big guy in that, von Braun was huge.

You must have known von Braun.
He was a very interesting guy. I met him at the Cape for the formal launch and

THIRD STAGE:
THE OTHER SIDE OF SKY

'There was a time when men stared longingly at the white patches on the globe and were fascinated by the challenging words "terra incognita". The longing led to action; all the lands were explored, all the seas were charted; the eternal silence of the poles was broken and now we live in a world almost without secrets.'
Leonard de Vries, *The Second Book of Experiments*

THE INTERNATIONAL SPACE STATION
– YOUR HOME AWAY FROM HOME

I make my way outside the European Columbus module, feeling its thin aluminium skin as I go, and into the Russian segment. I immediately notice a contrast in tone: the olive-green colour reminiscent of a different era of Russian space stations. Above the hatch a crucifix and pictures of Korolev, Tsiolkovsky, and of course Yuri Gagarin. The small folding crew dining table is down. Exploring the International Space Station mock-up at the European Astronaut Centre in Cologne, you get a pretty good sense of the geography, although in Cologne, unlike space, it has things like steps, floors and ceilings.

Low earth orbit (LEO) is anywhere from about 160 km and 2000 km above the earth. This is the realm of the Space Station. Up here, you will be falling around the earth at over 17,000 mph and orbiting every 90 minutes. 'International' is the key word to the success of this operation. Many nations, many languages. A common human goal: to explore the unexplored. To push ever further with our knowledge. The ISS has now been permanently crewed for seventeen years[*] – 227 astronauts from eighteen different countries. Where once individual nations raced for space, now we work together.

When you arrive on the ISS you'll be greeted by hugs from your international astronaut friends, a good meal will be waiting, calls to the family back on earth, tweets to your legions of Twitter followers, and through the Cupola, the glass-bottomed-boat viewing area, a spectacular panoramic view of planet earth suspended against the black backdrop of the cosmos awaits you, which you can instantly share with the 7.4 billion people below you, as well as reminding you why you're up there in the first place. In your new home there is no up or down. Every inch of wall space is taken up by storage, laptops, cables, pens, pencils, webbing, experiments and cameras. The cubbyhole sleep stations are situated around all four walls, just as Stanley Kubrick envisaged. Artificial light illuminates your world, and a 70-decibel racket of air-conditioning fans, ventilators and experiments will be the soundtrack for your stay. For six months or a year, this will be your world. By the airlock, the two white puffy American EMU spacesuits face each other like sentinels in a pharaoh's tomb, guarding the sacred door to another realm.

HOW TO JUMPSTART A DEAD
RUSSIAN SPACE STATION

The space stations of the 1970s, 1980s and 1990s were a far cry from the ISS. Basic and cramped and isolated from the earth. No Michelin-starred food, espresso machines and WiFi. The astronauts and cosmonauts

---

[*] October 2000 was the last time all humans were on the planet at the same time.

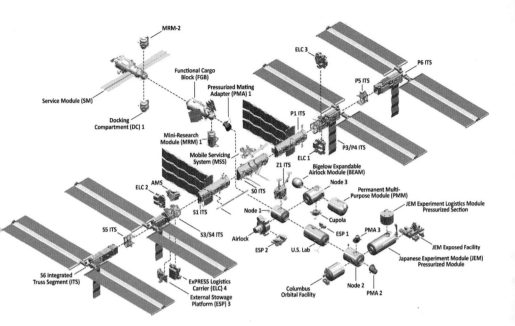

MRM-2

ELC 3

P6 ITS

Functional Cargo
Block (FGB)

Pressurized Mating
Adapter (PMA) 1

P5 ITS

Service Module (SM)

P1 ITS

Docking
Compartment (DC) 1

Mini-Research
Module (MRM) 1

P3/P4 ITS

Mobile Servicing
System (MSS)

ELC 1

Z1 ITS

Bigelow Expandable
Airlock Module (BEAM)

ELC 2

AMS

S0 ITS

Node 3

Permanent Multi-
Purpose Module (PMM)

JEM Experiment Logistics Module
Pressurized Section

S1 ITS

Node 1

Cupola

S5 ITS

S3/S4 ITS

Airlock

ESP 1

PMA 3

JEM Exposed Facility

ESP 2

U.S. Lab

Japanese Experiment Module (JEM)
Pressurized Module

S6 Integrated
Truss Segment (ITS)

ExPRESS Logistics
Carrier (ELC) 4

External Stowage
Platform (ESP) 3

Columbus
Orbital Facility

Node 2

PMA 2

who lived on them for months at a time were the true pioneers. Despite the hardship, cosmonaut Valentin Lebedev describes his life on the Russian space station Salyut 7 in 1982 with fondness:

I look around the station and view it with a different attitude. Now I think of it as home. The whole place looks so familiar. Everything in it is so near and dear to me now. When I look at the interior of the station, I feel no alienation, no sense that my surroundings are temporary or strange. Everything is ours. We've touched every square millimetre and object in here. We know exactly where every piece of equipment is mounted, not from documentation but from memory. Many little details, such as photographs on the panels, children's drawings, flowers and green plants in the garden [the Oasis, Fiton and other plant growth units], turn this high-tech complex into our warm and comfortable, if a little bit unusual, home.

On 2 October 1984, the Soyuz T-10 crew of Leonid Kizim, Vladimir Solovyov and Oleg Atkov prepared to leave Salyut 7. As they were shutting up shop, preparing the station to be temporarily mothballed in automatic mode, they left the customary crackers and salt on the table as a gift for the next crew to visit. The station would be empty for the time being: in hibernation, its systems would be controlled remotely from the earth. But after they left and returned to earth, a transmitter malfunction led to a short-circuiting of the electronics and all the electrical systems shut down. This was something that couldn't be fixed from the ground. The station was dead: comatose, orbiting the earth in suspended animation. With its solar panels no longer facing towards the sun, it began to freeze. A spare transmitter was on board, but it would need an electrician to install it. Abandoning the station completely and bringing it crashing back to the earth was one

Above: **The International Space Station**

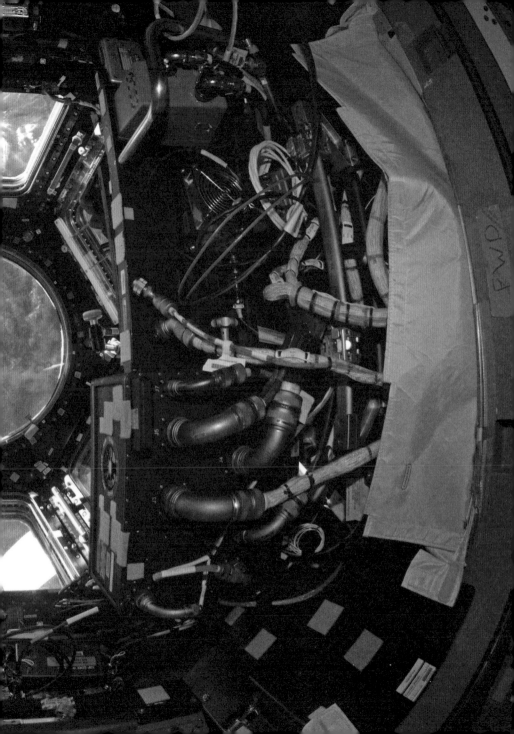

The first job was to attempt to equalize the pressure between the Soyuz spacecraft and the space station, which involved opening the ship's hatch and then a smaller porthole to equalize the pressure in the station airlock, beyond which was the station itself. A hissing meant the pressure was starting to equalize between the two crafts. Dzhanibekov could smell the stale air.

Not only was the station completely dead, it was seriously cold. Dressed in winter coats and woolly hats, they began the process of bringing it back to life. With no electrical power on board, the station was only illuminated by the sunlight every ninety minutes. When it was in the dark they used flashlights. The critical next job was to do a chemical test on the quality of the air, which seemed normal. Proceeding inside the station there was not a flicker of electricity. Spit would freeze to the walls and icicles hung from the pipes: 'It felt like being in an old, abandoned home. There was a deafening silence pressing upon our ears.' The crackers were waiting for them on the fold-down table. There was a danger of CO_2 build-up with two people on board. It was decided that one would wait in the Soyuz, keeping an eye on the CO_2 in the station.

With limited water supply, and suffering from physical exhaustion, it was a race against time to get the station back online. They needed power, but to jump-start the station from the Soyuz meant risking damaging it and becoming stranded. Instead they jury-rigged the station's dead batteries directly to the solar panels and used the Soyuz to rotate the station so the solar panels faced the sun once again. The station slowly began to come back to life. They fixed the water and air purification systems, and the faulty communication system problem was also diagnosed and fixed. 'That day was the first happy spark of hope in that mountain of problems, unknowns and hardships that Volodya and I were faced with solving.' Dzhanibekov remained on board Salyut 7 for 110 days.

option, but the political will to keep the space programme moving meant that a daring rescue was more suitable.

Docking a spacecraft with a dead frozen space station had never been attempted – it would be another 'space first', at a time when 'space firsts' were an important currency. The operation became what was described as 'one of the most impressive feats of in-space repairs in history'.

Vladimir Dzhanibekov and Viktor Savinykh were chosen for this extremely perilous mission, blasting off on 6 June 1985 from the Baikonur Cosmodrome on Soyuz T-13. The third Soyuz seat had been removed and filled with water and other equipment required for the repair. Night-vision goggles were packed in case the docking had to be done at night, and gas masks in case the air on board was unbreathable. Using hand-held laser guidance tools, they attempted the manual docking with Salyut 7 some four months after communication had been lost. But had the station completely depressurized? What was the temperature on board? If there was air, was it fit to breathe?

Above: **Viktor Savinykh and Vladimir Dzhanibekov**

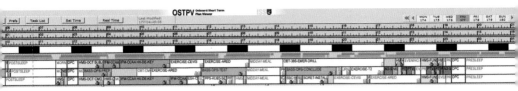

CONFLICT

'We got along together just fine.
We were bound by a common enemy:
Mission Control.'
William Pogue

Six weeks into the 84-day record-breaking
American Skylab 4 mission of 1973-4, rookie
astronauts Gerald Carr, William Pogue and
Edward Gibson had just about had enough.
They were running a punishing schedule of
medical and scientific experiments, as well as
solar and comet observations, photographing
the earth and a series of EVAs. As the heavy
workload built up, they started to slip behind
in the schedule and make mistakes in their
work. Tiredness, low morale and resentment
began to creep in. There had already been
minor friction with Mission Control over
the non-reporting of a bout of Pogue's space
sickness. Frustration over what the crew
perceived as an unrealistic work schedule
began to build, with things eventually coming
to a head when both the crew and ground
control aired their grievances to each other
over the radio. This was wildly exaggerated
in the popular press, with stories suggesting
the crew had deliberately turned their radios
off and refused to work in a full-blown act of
space mutiny. The ground crew was forced to
change the schedule and the incident threw
into question the importance of morale on long
duration space flights. The astronauts even got
into trouble after the mission, with a knuckle-
rapping over the inadvertent photographing
of the top secret military base known now
as Area 51 – the one and only place on earth

Above: **ISS daily planning software**

they weren't allowed to photograph. For all
three crew-members it was to be their only
space flight and marked the end of American
involvement in long duration space flight until
the joint Russian and American Shuttle–Mir
Program some twenty years later.

Being an astronaut on a space station
mission isn't about silver spacesuits, planting
flags and memorable first lines. Every minute
is accounted for, from early morning until
bedtime and the days are planned along a
strict visual timeline using special software
(Onboard Short Term Plan Viewer, OSTPV,
and later the Operations Planning Timeline
Integration System, OPTimIS).

As you plough across the timeline, the
work might become tiresome and tedious.
Resentment and conflict might begin to build.
Things might be fine for the first few weeks,
but the novelty and excitement wear off as the
weeks and months roll by and the grind begins.
'Put on your blinders and head North' was the
phrase borrowed from an Arctic explorer and
used by NASA psychologist Al Holland.

Over the forty-odd years of long duration
space flight, much has been learned about
what astronauts are capable of, as well as
how to manage the relationship with those on
the ground.

SHUTTLE-MIR AND THE THIRD MAN

Who is the third who walks always
 beside you?
When I count, there are only
 you and I together
But when I look ahead up the white road,
There is always another one
 walking beside you.
T. S. Eliot, *The Waste Land*

While the Americans were first on the moon, the Russians were gaining the lion's share of experience with long duration space flight, through their series of Salyut space stations and Mir, the last of which a group of seven Americans would inhabit on missions between 1994 and 1998.

One of those was Jerry Linenger, an all-American square-jawed naval officer. Things didn't go well. As well as feeling physically isolated, Linenger and the other American visitors had to deal with the deep cultural and institutional differences between the two nations. Like many in the Russian space programme, the Mir commander Vasily Tsibliyev was deeply superstitious and had consulted the famous Russian astrologer Tamara Globa, who had advised many of the cosmonauts what the stars held for them. She had foretold difficult times ahead. The Russian cosmonauts were under personal financial pressure to make sure things ran smoothly – a murky culture of financial incentives, bonuses and deductions for work carried out, which exacerbated a culture of half-truths and safety cover-ups. Linenger also found himself on board an ageing space station beset with leaks, breakdowns, broken oxygen generators and broken toilets, an environment that the Americans knew technically very little about, and with two cosmonauts who spoke very little English.

One evening, cosmonaut Aleksandr (Sasha) Lazutkin had changed one of the solid fuel oxygen generators (SFOG), which sparked, causing a blowtorch-like flame and billowing smoke. The fire was eventually brought under control but there were differences in reporting its severity and length. Was it just ninety seconds? Or a full fourteen minutes, as Jerry insisted? The Russians wanted to play down the severity and were ultimately in charge of safety on board the station, but the Americans thought differently, all of which fed a feeling of mutual mistrust.

Jerry became increasingly isolated, and frustrated. One day, about four months into the mission when things were at their worst, he was exercising on the treadmill when he was visited by an unknown crew-member. It was his father, who had passed away some years ago. He was standing there next to him, offering reassuring words of encouragement. The 'Third Man factor' is a psychological phenomenon where explorers, climbers and those under great stress and danger sense another's presence, often that of a missed loved one, and this presence is able to provide comfort, support or advice to those who feel like they are facing things at their worst. Jerry insisted:

It was my dad, it was him and I have no doubt that it was not an alien or anything else… It was my dad there to help me and it wasn't a scary thing – it was a wonderful thing to happen, and it helped me get through the end of that mission successfully.

The American baton was passed from Linenger to Michael Foale, the British American astronaut who replaced Linenger on Mir. He was aware of some of these problems and organized morale-boosting film nights for the crew, including his favourite films *2001: A Space Odyssey* and the great space drama *Total Recall*.

One incident came to define this mission: an experimental manual re-docking was to be attempted with a Progress supply spacecraft. Vasily Tsibliyev was doing this docking almost blind, using a new untested manual TORU system (Teleoperated Rendezvous Control System) that relied on visual information from a TV camera mounted on the Progress relayed to a monitor on the Mir – a cost-cutting measure. It had almost ended in disaster when the same manoeuvre was attempted during Linenger's mission, and indeed Linenger had warned Foale about it. The Progress ship was coming in too fast and Vasily was reliant on the others giving him visual cues as to where the Progress was from the Mir's small windows. Sasha Lazutkin and Michael desperately searched the ink-black sea for the approaching craft. Impact between the Progress and Mir was imminent. Sasha described seeing it approaching for the first time: 'It was full of menace – like a shark. I watched this black body covered in spots sliding past below me.'

Sasha barked at Michael, 'Майкл, на корабль!'*, ordering him to save his life in the Soyuz that was docked onto Mir. On impact, the hull of Mir's module 'Spektr' had been breached and the station was spinning out of control. Michael could feel his ears pop as the pressure inside Mir dropped, with the deafening sound of the master alarm. Death was minutes away. Michael waited in the Soyuz for the others to join him, but the resilient Russians had no thoughts of evacuation and soon, with Michael's help, they began to try to rescue the station.

They had minutes to seal off the breached Spektr module, first removing all the various cables that lay through the hatch. They tried to cut them with a small dining knife, before eventually managing to uncouple them, get them out of the way and seal the damaged module with one of the hatch covers – the pressure difference between the two modules sucking it into place.

The spin of the station caused by the impact had left Mir without power – its large solar panels no longer pointed at the sun. No power meant no communication with earth. Darkness descended, together with an overwhelming silence, punctuated only by the crackling of the ice on the hull as the cold began to wrap its fingers around the craft. As the station lost consciousness, the crew had a precious moment to pause and look outside to see the universe as it really is – a moment free from the distractions of human technology – black, star-filled and silent, illuminated from below by the feathered glow of the Auroras above the poles.

There was to be no abandoning of the ship. Michael made a plan to reorientate the station so the solar panels faced the sun again using the Soyuz thrusters, estimating the speed of the roll by using his thumb against the movement of the fixed stars and some mathematical calculations. It worked. The roll of Mir was halted and some basic power was returned. The astronauts were saved from death at least, although the power was still out in the Kvant-2 module that had the toilet and the crew had to urinate and defecate into some condoms and bags that the German astronaut Reinhold Ewald had left over from one of his experiments. Michael was forced to sleep in the airlock, 'like a dog'. Eventually, with Mir back online, Michael reinstated their cherished film night. On one particular evening, the three friends sat back and watched *Apollo 13*, with Foale providing the translation. A week later Jim Lovell, the commander of *Apollo 13* himself, told them on the radio that he thought they'd had a much worse time than they had. It was the perfect film choice to mark the end of what had been one of humanity's most fraught incidents in space.

After almost two decades without major incident, it's easy to take the ISS for granted. We're completely familiar with life on board. These days astronauts don't have to fight for survival. Instead Chris Hadfield serenades us on his guitar, and Tim Peake presents a Brit Award live from space wearing a tuxedo. With the constant rotation of international crews, the gourmet food and a constant presence on social media, astronauts have become part of our daily lives, dazzling us with their presence as well as a constant stream of ever more spectacular images of the earth. This miracle of our age didn't just happen by luck. We have the early pioneers who knew of no such luxuries, who roughed it out on board Skylab, Salyut and Mir, with all their trials and tribulations, to thank.

---

* 'Michael, to the ship!'

END OF LIFE
– SPACE DEBRIS AND THE RETURN TO EARTH

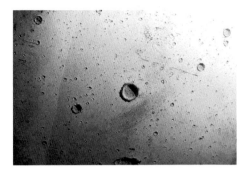

'Devills and wicked spirits, who, the first day of my arrivall, came about mee in great numbers, carrying the shapes and likenesse of men and women, wondring at mee like so many Birds about an Owle, and speaking divers kindes of Languages, which I understood not, til at last I did light upon them that spake very good Spanish, some Dutch and othersome Italian, for all these languages I understood.'
Francis Godwin, *The Man in the Moone*

In 1957, earth's second moon Sputnik 1 was the only man-made object in low earth orbit. These days it's getting quite crowded up there. In the intervening time, some 6600 satellites have been launched. These objects have been our eyes and ears in space for sixty years, revolutionizing life on earth. There are 3600 still up there, with fewer than one third still operational.

Space junk is becoming a big problem. More than 17,000 objects are currently being monitored by the US Space Surveillance Network, but beyond what can be seen and tracked, many millions of tiny flecks of paint, nuts and bolts, and even astronaut Piers Sellers's spatula that he dropped during an EVA, are all travelling so fast they can destroy a spacecraft, or puncture a spacesuit and the skin of an astronaut like a rifle bullet. The problem is only going to get worse as space access becomes cheaper, and satellites become smaller and more ubiquitous. With more and more space junk being constantly added, living and working in low earth orbit will become increasingly problematic. There are various exotic housekeeping schemes being planned to deal with it all, from orbiting collecting nets, to robotic arms and laser cannons.

One such object up there was a Russian Orlan spacesuit, known as 'SuitSat'. The ghostly suit had come to the end of its working life and was ejected from the ISS in 2006, cast adrift into space. Like the multilingual spirits that accompanied Domingo Gonsales through this realm, 'SuitSat' was a human-shaped broadcasting device with a radio transmitter on its helmet which sent greetings in five different languages to anyone who could tune in to its frequency, before eventually returning to earth as a shooting star.

At the end of life (EOL), the orbit of a satellite or space station will decay. This can happen either naturally over time, or deliberately using thrusters. When its altitude is lowered, its speed will be naturally slowed by atmospheric drag – the molecules of air at the top of the atmosphere hitting the object. As its altitude continues to decrease and the air becomes denser, the object's high velocity causes the air to compress beneath it, which in turn generates vast amounts of heat. This is the same process as when you touch the hot end of a bicycle pump after the tyres have been quickly inflated. Enough heat is produced for the object to burn up in a fireball or completely vaporize. This is why heat shields are required for spacecraft like Soyuz returning to earth.

Sometimes bits of debris, especially from a very large object like a space station, may make it past the furnace of re-entry to the ground. If you're really unlucky, you might

Above: **Pock-marked detail of recovered Salyut 7 helium tank**

get hit by something. Next time you're in the town of Manitowoc, Wisconsin, USA, head for 610 N Eighth Street. In the middle of the road you will see an embedded metal ring. This is the spot where a piece of the Soviet Sputnik 4 survived re-entry and fell to earth on 6 September 1962, an event that the town still celebrates in its annual sci-fi 'Sputnikfest'.

Most of the Salyut 7 space station burned up when it was finally brought down to earth in an uncontrolled re-entry across the skies over Argentina on 7 February 1991. But some fragments made it to the ground, much of the debris falling near the town of Capitán Bermúdez, including a cylindrical titanium helium tank and various fragments. Several of the pieces now reside at the local Oro Verde observatory. A sizeable chunk almost killed a local resident, according to an Australian newspaper:

A piece of the abandoned Soviet space station Salyut-7 crashed into a house in the town of Capitan Bermudez, about 300 km north-west of Buenos Aires, causing no injuries, a police spokesman said. National Atomic Energy Commission officials began examining a 'glowing washing-machine-sized' object late on Thursday which crashed into the yard outside the house, said its occupant, Dalia de Palazzo. The woman had been ironing clothes at 2pm Canberra time on Thursday when she had heard a 'frightening noise'. Outside she had found the piece of wreckage glowing in a crater in her patio.

THE SPACECRAFT CEMETERY.
THE OCEANIC POLE OF INACCESSIBILTY

'Ah, sir, live in the bosom of the waters! There alone is independence! There I recognize no masters! There I am free!'
Captain Nemo in *Twenty Thousand Leagues Under the Sea* by Jules Verne

23 March 2001 marked the end of the life of the Mir space station, most of which burned up in the atmosphere across a predetermined disposal corridor, but what remained ended up at a place called

Point Nemo in the middle of the southern Pacific named after Jules Verne's captain from *Twenty Thousand Leagues Under the Sea*. It's also known as the Oceanic Pole of Inaccessibility and has a place of special geographical significance, being the point on earth that's officially furthest from any land – the nearest solid ground is some 1650 miles away. If you find yourself there, you are officially a long way from anywhere. The remains of our flotilla of satellites and spacecraft now lie at the bottom of this most lonely of lonely spots. This is the Spacecraft Cemetery.

When the International Space Station reaches the end of its life, around the middle of the next decade, it will be de-orbited for a destructive re-entry. Like Mir, tons of metal will vaporize in the atmosphere like a great comet above our heads, but the fragments that remain will end up here, out of sight and out of mind. The flotsam and jetsam from low earth orbit resting for ever in a very different ocean, every bit as black and inhospitable.

Above: **The space debris problem**
© Agnes Meyer-Brandis, VG-Bild Kunst 2017

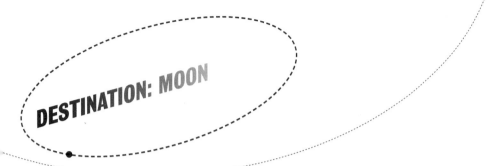

DESTINATION: MOON

'After the time I was once quite free from the attractive Beames of that tyrannous Loadstone the earth, I found the Ayre of one and the selfe same temper, without Winds, without Raine, without Mists, without Clouds, neither hot nor cold, but continually after one and the same tenor, most pleasant, milde, and comfortable, till my arrivall in that new World of the Moone.'

Francis Godwin, *The Man in the Moone*

TOURIST INFORMATION

Imagine that you're standing on the surface of the moon and that you're able to breathe somehow, unencumbered by bulky spacesuits, free to roam and explore at will. The universe around you starless and Bible black, the constellations washed out from the reflected light off the moon's surface. The pattern from the tread of your shoe imprints in the fine powdery dust. It behaves strangely when you kick it – with no atmosphere for the dust to be suspended in, the particles fall instantly back to the ground. Already your legs and clothes are stained grey. Around you are huge boulders, and moon stones small enough to put in your pocket. You try your first small step in the one-sixth gravity. You feel light, like walking across the tight skin of a trampoline, or buoyant – like walking on the bottom of a waterless ocean. The horizon is closer to you than it is back home. Tread respectfully. You are only a visitor here. A world of rilles, outcrops and mountains. Impact craters of every size, some too big to comprehend. Craters and craters within craters. Perhaps you have your roll call of the great and good they are named after: Tsiolkovsky, Goddard, Oberth, von Braun and von Kármán. And the great polar explorer Ernest Shackleton, at the vast South Pole–Aitken basin, where ancient water-ice lies locked in billion-year-old shadows. If you wanted to visit the Man on the Moon himself, you could travel to the Sea of Vapours – his nose – and the Sea of

Clouds – his mouth – and the Sea of Showers and the Sea of Serenity – his eyes. If you have your Michael van Langren moon map from 1645 to hand, you'll notice one of the craters is called *Gansii*, named after our heavy-lifting, migrating moon geese. Perhaps standing here you might look up at the light-blue marble earth and cover it with your outstretched thumb. And in doing so blot out the history of everyone who ever lived.

MOON FACTS
The moon is our first solid stepping stone in the ocean of space. Only three days away by rocket (if you want to get captured by its gravity; quicker if you want to just whizz on by), or nine days by geese with the moon on their wings. The moon has been our constant companion orbiting the earth for 4.51 ± 0.01 billion years, with its gravitational tug that causes our tides and maintains the earth's tilt – currently around 23.5° – that gives us our seasons. Our moon is unusually large for our planet. Not the biggest moon in the solar system, but by far the biggest in comparison to its home planet.

Where did it come from? Several ideas swirl about, but the front-runner currently is that during the early formation of the solar system a Mars-sized object called Theia*, the mother of mythical moon goddess Selene, slammed into the earth with a glancing blow, knocking a chunk of the earth into space and destroying itself in the process. An accretion disc of planetary debris then formed around the earth, with gravity eventually doing its thing, clumping it all together into our large moon.

Large as it is, the moon is too small for gravity to hold on to any atmosphere, which means, among other things, that meteorites that might otherwise burn up in our atmosphere hit its surface directly, giving it its pock-marked cratered look. A lack of atmosphere or any magnetic field means the radiation from the sun bombards the surface directly. With no weathering or current major geological activity, the moon's surface seems unchanging. The darker patches, the *maria* or seas, are cooled basaltic lava flows, so we know the moon had active volcanism in its history. Far from being a lifeless sterile rock, the moon holds the secrets of the formation of our solar system, the promise of water and raw materials, as well as a deep emotional pull.

The moon is (on average) 238,856.3 (±) miles from earth, and is moving away from us at about 3.8 cm per year. That's about the speed of your fingernail growth. We know this thanks to the cat's-eyes-like retro-reflector grids installed there by three of the six Apollo missions, and the Soviet Lunokhod rovers that are still parked on the surface – you fire a laser beam at them from earth and measure the time it takes for the light to bounce back, giving you the exact distance. The moon is 1.2 light seconds away. By coincidence, the diameter of the 'corner mirrors' on these reflectors is also 3.8 cm.

A much more staggering coincidence is that the size of the moon and the size of the sun, as we perceive them as discs in the sky here on earth, are identical – the sun's diameter is 400 times bigger than the moon, but it is also 400 times further away, hence they look exactly the same size, and it is the reason we have total eclipses when the moon passes in front of the sun. As the moon is slowly receding from us, in a few million years this glorious phenomenon will no longer be there. But then again, neither will we.

---

* It wasn't called Theia at the time. Names didn't catch on until people came along several billion years later.

GETTING THERE – SATURN 5 AND N1

'… if you drive a car with an automatic transmission and you come to a red light you stop – you put your foot on the brake. Light turns green. You don't put your foot on the accelerator, you just take your foot off the brake what happens? You start moving very, very slowly right? That's exactly the way we launched… the same kind of motion, the same kind of acceleration…'
Al Worden

If you want to go to the moon (call it a ten-day round trip minimum), you'll need a vehicle that can do some heavy lifting; once in space, you'll need a spacecraft to get you and your stuff to the moon and back. And you'll need a lander to get you to and from the moon's surface. A flag to plant once you're there. A way to get you safely back through the earth's atmosphere. And fuel for all legs of the trip. That's a lot of kit to get off the ground. In the 1960s and 1970s the Americans used the famous three-stage Saturn V designed by von Braun: standing 363 feet high and weighing 2.8 million kg, it remains the most powerful rocket ever to fly, soon to be surpassed by the planned SLS.

In the political race to the moon, the Soviets also had a heavy launch vehicle, the mighty (yet secret until the Soviet Union started to crumble) N1 rocket stack, with its first-stage rings of 30 NK-33 engines which could carry the Soyuz/Zond family of spacecraft and the single-occupant LK lunar lander. The towering four-stage N1 rocket never made it into space. All four unmanned test launches ended in failure; the second attempt on 3 July 1969 exploded seconds after lift-off at the Baikonur Cosmodrome, igniting tons of propellant and creating one of the biggest non-nuclear explosions ever. The N1 programme was cancelled in 1974, long after the race had been won by the Americans. The only N1 rockets you will see now are in scrap-metal form. No doubt some small parts still litter the Kazakh steppe, some of the bigger panels are converted into shelters on the Cosmodrome, and some recycled as playground furniture in the town of Baikonur itself.

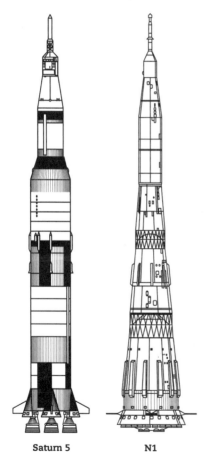

| Saturn 5 | N1 |

Opposite above: **'The Person who doesn't collect scrap metal won't be getting to the moon'**
Opposite below: **Pieces of N1 rocket in a playground**

FINDING YOUR WAY - KNIT YOUR OWN SATNAV

'Star patterns are very important. Through the atmosphere you can see about ten million stars. We had a sextant on board that was exactly the same as the ancient mariners use. I did a lot of star sightings on the way out. What we had to do was a sighting between two stars, and we'd record that angle, and then another pair of stars, and then another pair of stars, and the computer would calculate all that and it would give us our attitude with respect to the universe...'
Al Worden

When travelling in space, knowing where you are and where you're going is important. Domingo Gonsales's geese instinctively knew how to get to the moon. Al Worden and the other twenty-three lunar travellers had no such luxury. Aiming for the moon, a moving target 250,000 miles away, in a rocket launched from another moving object requires millions of calculations and corrections. As well as a sextant, also on board were a gyroscope and accelerometer-based inertial guidance navigation system developed by MIT.

Just as important was a computer. Computers need written software instructions to tell them what to do and memory to store the information. The Apollo Guidance Computer (AGC) used a forgotten technology called 'rope memory'. Each 1 and 0 of software binary code was an individual physical copper thread. They were hand-woven using a special needle and a loom around small doughnut-shaped magnetic cores by a group of skilled women, working at the defence company Raytheon who manufactured the computer, in a process known as the LOL method (little old lady). A wire through the core would be a 'one', and around the core a 'zero'. The information would be within the pattern of the weaving itself. Painstaking work, and a nightmare if you make a mistake. That gave the computer around 69 kilobytes of 'Read Only Memory'. In other words, not very much. In a world where computers were the size of rooms, the AGC – a computer that could fit in a tiny spacecraft – was arguably the most challenging bit of engineering of the entire Apollo project, despite only having the memory of a musical greeting card.

SPACE AGE needleworker "weaves" core rope memory for guidance computers used in Apollo missions. Memory module will permanently store mission profile data on which critical maneuvers in space are based. Core rope memories are fabricated by passing needle-like, hollow rod containing a length of fine wire through cores in the module frame. Module frame is moved automatically by computer controlled machinery to position proper cores for weaving operation. Apollo guidance computer and associated display keyboard are produced at Raytheon Company plant in Waltham, Massachusetts.

WHO, WHAT, WHEN AND WHY OF APOLLO MISSIONS

MISSION	LAUNCH DATE	CREW	NOTES	MISSION DURATION
Apollo 1		Virgil 'Gus' Grissom, Edward White, Roger Chaffee	Launch test fire resulted in tragedy.	
Apollo 4 (after re-ordering)	9 November 1967	Unmanned	Maiden flight of the Saturn V rocket. Orbital tests.	
Apollo 5	22 January 1968	Unmanned	Orbital tests.	
Apollo 6	4 April 1968	Unmanned	Orbital tests.	
Apollo 7	11 October 1968	Walter Schirra, Donn Eisele, Walter Cunningham	First three person orbital test mission. • First live TV broadcast from space. • First astronauts to have colds in space: 'We're up to our asses in used tissues' (Cunningham).	10 days, 20 hrs, 9 mins, 3 secs
Apollo 8	21 December 1968	Frank Borman, Jim Lovell, William Anders	First non-fictional humans to fly to, and orbit, the moon and return. • Flew with a clandestine miniature bottle of Christmas brandy. • Frank Borman becomes first human to vomit in space. • Famous 'Earth Rise' photograph of Earth from the moon taken by Anders. • First humans to see the far side of the moon.	6 days, 3 hrs, 42 mins
Apollo 9	3 March 1969	James McDivett, David Scott, Russell 'Rusty' Schweickart	Dubbed 'A connoisseur's mission' due to its success (Schweickart). • Docking and other hardware testing. • A Command Module EVA performed by Scott.	10 days, 1 hr, 54 secs
Apollo 10	18 May 1969	Thomas Stafford, John Young, Eugene 'Gene' Cernan	Lunar landing 'dress rehearsal'. • First crew to shave in space. • 'Fly Me To The Moon' (Howard/Ballard) song played on board.	8 days, 3 mins, 23 secs
Apollo 11	16 July 1969	Neil Armstrong, Edwin Aldrin, Michael Collins	First moon landing. • First words spoken: 'That's one small step for (a) man...' (Armstrong). The vital 'a' got lost in the broadcast. • 'Magnificent desolation' (Aldrin). • Lunar EVA duration: 2 hrs, 31 mins, 40 secs.	8 days, 3 hrs, 18 mins, 35 secs
Apollo 12	14 November 1969	Charles 'Pete' Conrad, Richard F. Gordon, Alan L. Bean	Lightning strike on launch. • Astronauts recover parts of 'Surveyor 3' lander from moon's surface. Streptococcus mitis bacteria found alive on its camera after over two years on moon.	10 days, 4 hrs, 36 mins, 24 secs
Apollo 13	11 April 1970	Jim Lovell, Jack Swigert, Fred Haise	There was a problem (oxygen tank rupture) which was famously reported to Houston, and often misquoted. The actual phrase was in the past tense: 'Houston, we've had a problem.' Mission dubbed: 'a successful failure'.	5 days, 22 hrs, 54 mins, 41 secs
Apollo 14	31 January 1971	Alan Shepard, Stuart Roosa, Edgar Mitchell	Over nine hours of Lunar EVAs. • Golf balls hit on moon (six iron) by Shepard. • 'Psychic experiments' performed en route to the moon by Mitchell using a special deck of cards.	9 days, 1 min, 58 secs
Apollo 15	26 September 1971	David Scott, James Irwin, Alfred 'Al' Worden	First use of the lunar rover. • 'Fallen Astronaut' memorial left on lunar surface. • Falling objects in gravity fields demonstration performed by Scott by dropping a hammer and falcon feather. • First deep space EVA performed by Worden.	12 days, 7 hrs, 11 mins, 53 secs
Apollo 16	16 April 1972	John Young, Ken Mattingly, Charles Duke	Over 20 hrs of lunar EVA. Rover driven nearly 27 km. • Moon flatulence reported by Young, suspected caused by high citrus fruit diet. • Charles Duke leaves a family photograph on moon's surface.	11 days, 1 hr, 51 mins, 5 secs
Apollo 17	7 December 1972	Eugene 'Gene' Cernan, Ronald Evans, Harrison Schmitt	Harrison Schmitt becomes the first scientist/geologist in space. • Gene Cernan becomes the last man to walk on the moon.	12 days, 13 hrs, 51 mins, 59 secs
Apollo 18–20			Cancelled.	

MAN IN THE MOONE, FRAU IM MOND

For the 108 billion humans who have ever looked at it, the moon is a great cultural Rorschach inkblot test in the sky, a celestial blank canvas for all our human concerns through the ages: scientific, artistic and religious. In 1638 Domingo Gonsales took with him on that imaginary voyage a discussion of the new Copernican science – Space 1.0 – as well as the religious hand baggage of centuries past. En route, while pondering the latest ideas of gravity and magnetism, Gonsales's visions of devils and spirits in low earth orbit were a metaphorical challenge to the emerging scientific enlightenment. The first thing he did when greeting the inhabitants of the moon (after having a quick snack) was to cross himself, proclaiming '*Iesus Maria*', resulting in them falling to their knees. Christianity, like the ether, was seemingly all pervasive across the universe.

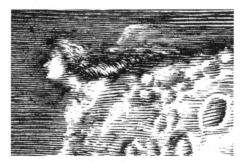

Continuing the theme a few hundred years later, on Christmas Eve 1968 the crew of Apollo 8 read out loud from circumlunar space the first ten verses of the Book of Genesis to the largest television audience of all time, which generated a legal challenge by atheist Madalyn Murray O'Hair, citing that it was a clear violation of the First Amendment's separation of religious freedom and the role of the government.* Buzz Aldrin also took Holy Communion before that first lunar EVA – in his astronaut's Personal Preference Kit he had packed a communion wafer, wine and a printed card from which he read: 'I am the vine, you are the branches. Whoever remains in me, and I in him, will bear much fruit; for you can do nothing without me.' (John 15:5.)

Above: **Cassini's map of the moon. Can you find the secret Moon Maiden?**

---

* antidisestablishmentarianism.

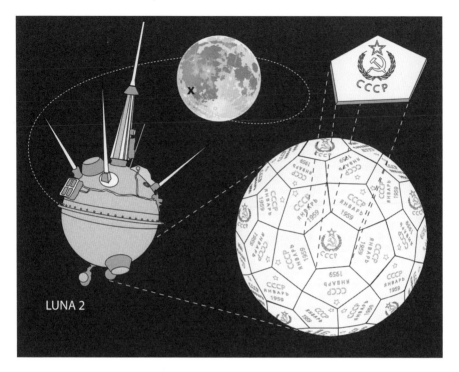

LUNA 2

Our science and folklore have always walked hand in hand. When we look at the moon, like the astrological pictures we see in the stars, the brain's highly developed pattern recognition software picks out pictures, and faces – a wide-eyed face on the mottled lunar surface, the man in the moon, caused by the variations of light and dark of the lunar highlands and 'seas'. But he's not alone – there's also a rabbit on the moon, called Yutu, favoured by the Chinese. In 1679, the astronomer Giovanni Cassini drew the most detailed, scientifically accurate map of the moon using telescopic observations, and managed to smuggle into one of its craters a portrait of his wife Geneviève, much as a writer might smuggle a secret message past his editor, hidden somewhere deep in the text addressed to a significant other.

TREAD LIGHTLY

There's a lot of our space hardware still lying on the moon. The Soviet Luna 1 and Luna 2 were the first space probes to escape the earth's gravity and make the 'cislunar' crossing, with Luna 2 crashing into the surface on 14 September 1959 like Georges Méliès's painful-looking projectile from the early film *Le Voyage Dans La Lune*. It carried with it (as well as a package of scientific instruments) two strange liquid-filled, explosive football-like devices that would detonate on impact, dispersing their 'CCCP'-engraved pentagonal panels like seeds. To what extent these have survived is a mystery, but if you're heading that way it would be worth a look.

Above: **Luna 2**

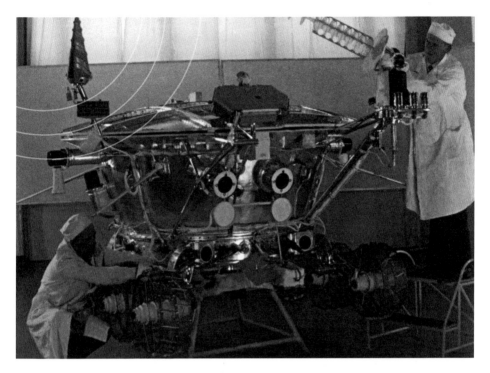

Or what about a visit to the six Apollo landing sites, to see first-hand the footprints and the flags? The sites themselves and the larger pieces of hardware are visible from the Lunar Reconnaissance Orbiter (LRO), which is right now orbiting the moon taking photographs. Imagine in the distance seeing the familiar lunar lander platform wrapped in its glinting insulating material like a golden fleece. Nearby you would see the lunar rovers from the Apollo J missions (15, 16 and 17) and the tracks they made from wheels made from piano wire. Gene Cernan's rover at Taurus–Littrow has a broken mudguard repaired with duct tape. How is it holding up? Those alien-looking remote Russian Lunokhod rovers are still up there. Lunokhod 2, which travelled 39 km across the moon's surface, is owned by private astronaut Richard Garriott in a legal first – he bought it for $68,500 in a New York Sotheby's auction in December 1993, and a credit card receipt was his only physical token of the purchase.

As with any archaeological site, pictures are the only things you can take. These precious sites need to be protected: from overenthusiastic visitors disturbing Neil Armstrong's first footprints, for example. In 2011 NASA published a document, 'NASA's Recommendations to Space-Faring Entities: How to Protect and Preserve the Historic and Scientific Value of U.S. Government Lunar Artifacts', to protect these places from people like you, and especially the robotic visitors from Google's inducement competition, the Lunar XPRIZE, who may get there very soon.

Above: **Lunokhod rover**

LUNAR XPRIZE

The Lunar XPRIZE is a competition for teams to complete a series of tasks by the end of 2017, a deadline that has already been extended and may be again. The rules are simple:

- Successfully place a spacecraft on the moon's surface
- Travel 500 metres
- Transmit high-definition video and images back to earth.

The prize is $20 million for the winner and two $5 million prizes for the runners-up and a bonus prize. There are international teams, and teams from countries including India, Israel, Malaysia and Brazil. All these teams have slick presentation videos and bold, inspiring mission statements: 'We choose to go to the moon not because it's easy, but because it's a good business,' claims Naveen Jain, the Moon Express chairman and visionary, to the *New York Times*. And there is a web series, 'Moon Shot', made by well-known sci-fi director J. J. Abrams, which you can watch on the website.

As you would expect, some are closer to the moon than others. Some teams have been absorbed by others, and some have pulled out. The American Moon Express team have approval from the Federal Aviation Administration and a launch contract. The German 'Part-Time Scientists' are heading to the Apollo 17 site, and have space on their Alina spacecraft to carry your keepsake or experiment for a few hundred thousand dollars a kilo.

As long as you can avoid disturbing the historical integrity of the various locations and objects referred to in the document, there is still some very useful science to be done, for example, with the laser retro reflectors left by Apollo 11, 14 and 15 still used for gathering data about the distance to the moon. Do not touch these. Especially the actual corner cubes, which bounce the laser beam back in the direction it came. You can make some careful field notes as to their condition and take some close-up pictures. Long-term degradation of optical devices on the moon has been observed and scientists are keen to know why. Solar radiation? Dust on the surface? Information about the degradation of man-made materials on the lunar surface would be of great interest. What about micrometeoroids? All this would be vital information to material scientists designing future space missions.

Apollo 15's Dave Scott and Jim Irwin left 189 objects up there, including the hammer and falcon feather used to demonstrate 'falling objects in gravity fields'. How have fifty years on the moon's surface affected a biological specimen?

Above: **Apollo 15's hammer and feather**

PART NUMBER	NOMENCLATURE	QUANTITY
SEB3100040-307	Camera, Hasselblad	1
SEB3100048-305	Lens, 60 mm	2
SEB33100291-301	Tether, EVA retractable (CDR)	1
SEB33100402-301	Brush, Lens	2
14-0228	Wet Wipes, Facial	Used of 9
14-0121-301 (2,3)	Food Assembly LM	Used of 3
SEB33100284-302	Lens, 500 mm	1
BW1060-003	Jacket, Assembly, ICG	2
BW1061-002	Trouser Assembly, ICG	2
209-42999-8	Lunar Roving Vehicle (LRV)	1
14-0112-01	Towels, LM utility (Red)	2
14-0112-03	Towels, LM utility (Blue)	2
SEB39100319-301	Hammer	1
LSC 340-201-529-8	Urine collection assembly, (small)	2
LSC340-1000-19-1	Urine receptacle system	1
N/A	Tools ALSEP deployment	1
SEB39105177-301	Flag Kit, Lunar Surface	1
A6L-400000-16	Garment , Liquid cooled	2
A7L8106062-03/04	Boots, Lunar-pr	2
SEB39106380-303	Rake, Lunar Sampling	1
SEB33100255	Camera/Power Pack Assembly, 16 mm L.S.	1
SEB33100794-301	Helmet and LEVA Interim Stowage Container Assembly	1
V36-601012-351	Sleeping Restraint Assembly	1
2265840-501	CTV Camera, colour TV	1

Here's a list of some of the things left by the Apollo 15 crew, including the part numbers for your reference. It's a useful list of what to pack when you embark on your lunar space adventure.

The NASA document also has a list of various 'science opportunities' to be carried out by would-be visitors, including 'Push biggest possible rock over edge of crater or rille', in the first instance to consider the 'geotechnical properties of the surface layer', but also as the author notes 'because it would be fun to push a big rock over a cliff'.

Over the years we have left tonnes of stuff on the moon. As well as the lunar rovers (Apollo 15, 16, 17), descent stages, and ALSEP (Apollo Lunar Surface Experiments Package), hundreds of objects were left on the surface to save weight and space for the return journeys of craft laden with precious moon rocks. Bits of spacesuits, Hasselblad cameras and lenses, food, soap, nail clippers, astronaut faeces and a lot of wet wipes. There are two golf balls up there too, struck by Alan Shepard.[*]

---

[*] If you go there, one is in 'Javelin crater', named after Ed Mitchell (Apollo 14) threw a pole from an experiment into it.

ART ON THE MOON

'My Homo sapiens, my cybernetic man, has in whole or part come true... I am optimistic about mankind. I think the future of man is out there... In twenty years, I think we'll be off to other solar systems.'
Paul Van Hoeydonck

There's more than one man on the moon. At 167:43:36 hours into the Apollo 15 mission, in a moment of quiet reflection, Commander Dave Scott placed a palm-sized solid aluminium stylized figurine of an astronaut in the lunar soil. Next to this, he placed a small plaque bearing the names of the fourteen men who had lost their lives in the pursuit of space travel up to that point.* At that moment the moon became the 'largest exhibition space in the known universe'.

The artist who had created the aluminium astronaut was a Belgian named Paul Van Hoeydonck, whom Dave Scott and Jim Irwin had met over dinner and then made an informal arrangement to place it on the moon. But things soured, as the significance of such an act differed between the two men. For Van Hoeydonck, this was a bold artistic statement – a grand vision shared by Tsiolkovsky, Musk and many others since, of humanity stepping beyond the cradle of the earth and into the cosmos at last. A symbol of triumph. For Scott though, the gesture was simply intended as a private reflective moment, a dedication to lost comrades. It became known as *The Fallen Astronaut*, not the name or indeed sentiment that the artist had intended.

The memorial was kept secret until the official press conference where Scott briefly mentioned it, with no mention of Van Hoeydonck's name – as had been informally agreed, the artist would remain anonymous,

at least for an extended period of time, refraining from any sort of self-promotion. But resentment soon crept in. Van Hoeydonck began to feel that he wanted credit for his work. He was further annoyed when he was later approached by Scott to make two copies of the statuette at the request of the National Air and Space Museum of the Smithsonian Institution in Washington, who wanted one to display. Not knowing his identity, they had referred to him in correspondence with Scott as the 'workman' which riled him intensely.

Paul Van Hoeydonck did make two replicas, one for display at the Smithsonian and another for the King of Belgium. His authorship finally became known to the world on a live TV appearance with Walter Cronkite, during the televised launch of Apollo 16. Plans were subsequently made by Van Hoeydonck and his gallery to produce a limited edition of 950 of *The Fallen Astronaut* for sale at $750 a pop, which resulted in an official investigation and was soon halted. Today, there are various reproductions and forgeries of *The Fallen Astronaut* that turn up for sale.

The figure represents two different artistic statements that somehow got lost in translation on the way to the moon – one about the future, the other about the past. If you look up at the moon at the foot of the Apennine mountains, and squint hard, you might just see it glinting.

---

* Missing from the list were Valentin Bondarenko and Grigori Nelyubov, two cosmonauts whose deaths at that time were unknown outside the Soviet Union. Perhaps, if you go back, you could carefully add a plaque with their names and the names of the men and women who have died since.

BASSETT, CHARLES A. II
BELYAYEV, PAVEL I.
CHAFFEE, ROGER B.
DOBROVOLSKY, GEORGI T.
FREEMAN, THEODORE C.
GAGARIN, YURI A.
GIVENS, EDWARD G., Jr.
GRISSOM, VIRGIL I.
KOMAROV, VLADIMIR M.
PATSAYEV, VIKTOR I.
SEE, ELLIOT M., Jr.
VOLKOV, VLADISLAV N.
WHITE, EDWARD H., II
WILLIAMS, CLIFTON C., Jr.

ANDY WARHOL'S TINY PENIS

The Apollo 11 moon landing was to become one of the defining events that shaped the aesthetic landscape of America at the end of the 1960s. Artist Robert Rauschenberg had been invited by NASA to attend the Apollo 11 launch, and his reflections of the event formed his *Stoned Moon* series of space inspired lithographs. The American space program wasn't just a science and engineering project – it was *the* cultural event of our age.

Paul Van Hoeydonck's *The Fallen Astronaut* might not be the only art up there on the lunar surface. In 1969, the doodles of six American artists were collected together in a single piece called *Moon Museum*. The artists were Rauschenberg, who just drew a straight line; David Novros; John Chamberlain; Claes Oldenburg, who drew a sort of Mickey Mouse; Forrest Myers; and Andy Warhol, who hastily scribbled a penis. This bonkers example of late-twentieth-century American contemporary art was miniaturized on a tiny ceramic tile, 1.4 cm by 1.9 cm, by technicians at Bell Laboratories. A handful of these were produced – one is still exhibited in galleries around the world – but the idea was to try to get one up to the moon. After official permission to get *Moon Museum*, on board the Apollo 12 mission was refused, it was allegedly smuggled in between the multi-layered gold-foil insulation blankets on the Apollo 12 Lunar Module descent stage (the bottom bit of the lunar module that remains on the moon) by an anonymous engineer from Grumman, the company that manufactured the strange spidery-shaped spacecraft.

But did it happen? The Apollo 12 commander, Alan Bean, himself now a painter, points out the obvious flaw in the story: why would an engineer risk everything to do such a thing on such a sensitive project? The mystery was investigated in the popular American PBS documentary series *History Detectives*. A telegram exists, dated correctly and reportedly from the mystery engineer who was in on the scheme and went by the pseudonym 'John F', which confirms that the artwork had indeed been smuggled on board. Interestingly, the launch-pad foreman at Grumman at the time, Richard Kupczyk, has since confirmed that small personal keepsakes, such as family photographs and mementos, had been surreptitiously slipped between the foil insulating layers by the engineers who had built the spacecraft, and although technically illegal, a blind eye had been turned as it didn't compromise the safety of the mission.

An art mystery that can only be solved by gently peeling back the layers of gold foil and having a very careful look.

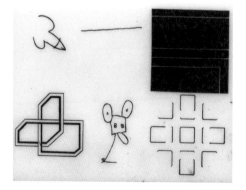

Left: **Moon Museum (various artists), 1969**
Opposite: **Apollo 12 Lunar Module**

BACK TO THE MOON

With small movements on the oversensitive control panel, I clumsily head through the door of the shelter and out onto the Shackleton crater at the moon's southern pole. A world of ice at the lunar pole of inaccessibility. I seem to be hovering a giant leap above the lunar surface, but manage to master the thumb-stick control that lets me stand. It is breathtaking. Magnificent Desolation.* I can see the flat plain of the crater rise up to a rim a few miles away. Our dome, like a grey igloo, is the only man-made feature I can see. The brightness of the grey lunar soil, in stark contrast to the black sky, is striking. I can make no footprints in it though. Like those twelve moonwalkers, I practise my gait, slowly at first and then, gaining confidence, I start to explore. The horizon is much closer than on earth. I pause just to look around. My limbs are invisible even though I can feel them. I hold my arms out but there's nothing there. The bright sun hangs in the black sky. At least I think it's the sun – could it be the washed-out earth? The engineer who's standing right beside me in an alternate reality tells me it's the sun. He tells me that all the detail I'm seeing has come from real lunar satellite data. Today, I walked on the moon for the first time. I can switch my imagination off. Idle fantasy gives way to experience.

I remove the headset, gobsmacked at the clarity and sophistication of the experience. Here, in a small computer lab deep in the European Astronaut Centre, they are developing a virtual reality system to train astronauts to explore the moon's surface. The charismatic director of the European Space Agency, Jan Woerner, has spoken much about the construction of a new 'moon village' – arguing the case for a truly international collective of robots and humans, space agencies and entrepreneurs working together on the surface. A focus for science and exploration. One could imagine a research station on the moon like the Concordia base in Antarctica. The lunar regolith itself could be used as a building material – 3D-printed structures designed to protect against radiation, built around inflatable templates. At ESA they're already working with home-made lunar material to do just that.

A lunar analogue facility will be developed right here in Cologne. A test bed for our return to the moon. This is where a virtual reality will pave the way for actual reality: a rehearsal space for a permanent presence on our nearest neighbour.

And this time we mean business.

'The moon's riches of gold, should they exist, ought to be placed in the hands of businessmen and not into those of visionaries and idealists.'
Frau im Mond

How much is the moon worth? As our planet's resources come under increasing pressure, should we be looking to the moon? Could it be mined? And if so, for what? Who would benefit financially? The moon is rich in natural resources, as well as being a celestial catcher's mitt for the metals and minerals that slam into it in the form of asteroids. Increasingly attractive are the moon's rare earth metals, which our earthbound high-tech digital industries rely on.

One of the moon's most discussed resources is Helium 3 – a helium isotope that is rare on earth, where it is filtered out by the atmosphere, but abundant in the lunar regolith where it is collected from the sun. Helium 3 might be an important source of fuel for nuclear fusion, the much discussed potentially game-changing energy solution. Helium 3 would provide a much cleaner fuel, but as yet workable fusion reactors coming online are still a long way off.**

---

* Aldrin's first words on the moon.

** Forever twenty years away, as the joke goes.

Perhaps the most important resource on the moon is water, locked up as ice at the poles as well as in the lunar soil, which could be extracted and converted into breathable oxygen and hydrogen to make rocket fuel.

Naveen Jain, the founder of the proto-commercial space company Moon Express, believes the next set of global superpowers are likely to be entrepreneurs exploring the moon for its natural resources to 'solve the world's biggest problems'. The idea of commercial exploitation of the moon is controversial, not least because of the Outer Space Treaty that prevents nations claiming sovereignty of the moon and any other celestial bodies. But as we look further out beyond the earth, our exploration is going to be increasingly driven by commercial ventures. Legal frameworks are already being drawn up to help such companies navigate these tricky waters.

STONED MOON

You can always visit the moon without leaving earth. In the Science Museum in London, for example, is a piece of the Apollo 15 'Great Scott' moon rock, part of a large sample taken by Dave Scott and Jim Irwin's mission.

Moon rock is spectacularly rare and considerably more valuable than gold, given the difficulty and expense of the delivery costs. How can you put a value on it? During the trial of Thad Roberts, a NASA employee who with three others stole 101 g of lunar and other samples from a safe in the Johnson Space Center,* a federal court valued the recovered haul at $50,800 per gram, a value in 1970s money based on the $110 billion cost of sending twenty-four men to bring it back to earth.

The very small amounts of moon rock on earth have got here in three ways. First, there are the lunar meteorites – chunks that naturally but rarely make their way to earth, blasted off the lunar surface, whizzing through the atmosphere and hitting the ground on earth. Even more unlikely is one of us chancing upon them, given that they look unremarkable and like any other piece of basalt. The trick is to go to a desert, where lumps of grey rock naturally contrast with a light-coloured sandy backdrop. Or go to Antarctica where they're very visible on the white canvas of snow and ice. So far we have discovered about 48 kg of lunar material that has naturally made its way here.

The second origin of lunar rocks on earth is the 382 kg hand-delivered by the men who visited the moon. Most of the samples of moon rock are stored in high purity nitrogen in building 31N of the Johnson Space Center – a state-of-the-art facility designed to withstand hurricanes, flooding and people like Thad Roberts breaking in. If you're a scientist who studies lunar rocks for research, you can apply to visit. For the rest of us there's an online virtual tour. Fifty-two kilograms are stored in a special vault at White Sands Test Facility, moved there in 2002 from Brooks Air Force Base in San Antonio, Texas where they had been stored since 1976. Trucks were loaded with accelerometers to record the vibrations during the journey. Hurricane Isidore, which hit the US at that time, was carefully tracked by Mission Control, such is the value and importance of these samples.

Our third source is the Soviet Luna missions, which returned 326 grams, 0.2 grams of which have constituted the only legal sale of moon rock to date, when three seed-sized pieces were sold at Sotheby's in New York in 1993, in their Russian space sale.

NASA keeps very strict records of all of their Apollo samples. Some pieces of the moon have been given as gifts: in 2004 the Mercury, Gemini and Apollo astronauts were given pieces, as well as the broadcaster Walter Cronkite. Other pieces were from both Apollo 11 and 17 missions (sample 70017), known

---

* Apparently Thad spent an out-of-this-world night with his girlfriend in a hotel room with the rocks spread about the bed.

23

as the 'Goodwill' rocks, given as gifts to each of the 50 states as well as to 135 countries. Once given, these presentation samples, embedded in a solid plastic dome and affixed on a wooden plaque with a legend and the country's flag, became the property of that country and outside NASA jurisdiction. It's worth having a look to see where your country's moon rocks are.* What we do know is that many of these gifts have vanished without trace over the years.

'Special Agent' Joseph Gutheinz is a lawyer, lecturer, undercover NASA cop and space detective. He's made it his mission to track down the missing rocks, including a sting operation called 'Lunar Eclipse' to try to find those from Honduras, which appeared on the black market, as well as getting his students at the University of Phoenix to trace the 158 others that have been lost, misplaced, stolen, given as illegal gifts, or simply vanished into thin air. The Arkansas Goodwill rock from Apollo 17 that had gone missing turned up by chance in an archive box of Bill Clinton's papers.

In the suburbs of Dublin is an old landfill site containing some 4 million tons of waste. Back in 1977, the Dunsink Observatory across the road caught fire and the remains were thrown into it. It's thought that Ireland's priceless moon rock gift from Apollo 11 was among this debris. A tough treasure hunt, but it could be well worth your while if you fancy having a root around.

HUNT THE 'MISSING' MOON ROCK
If you're in the UK you can visit our Goodwill Apollo 17 moon rock in the Natural History Museum in London. The UK's Apollo 11 lunar sample is safe and sound in a room at 10 Downing Street, sitting in plain view on a desk. I won't tell you which room or which desk, but if you go online and take the official 10 Downing Street virtual tour, it's in plain view. Happy hunting. Here's the link: www.gov.

uk/government/history/10-downing-street.
Or indeed, next time you find yourself at number 10, have a look around for yourself.

MOON VEXILLOLOGY

'As for the Yankees, they had no other ambition than to take possession of this new continent of the sky, and to plant upon the summit of its highest elevation the star-spangled banner of the United States of America.'
Jules Verne, *From the Earth to the Moon*

If you're thinking about conquering new worlds on your trip, you're going to need a flag to show you got there first. So much has been written about every technical aspect of the Apollo hardware, and yet the engineering story behind the moon flags themselves remains largely ignored. The planting of the flag was the cherry on the cake for the entire Apollo endeavour, yet the flag, which was to be seen by the entire world, was almost overlooked in the wild rush to get to the moon first.

The Committee on Symbolic Activities for the First Lunar Landing, set up only five months before the Apollo 11 launch, suggested a freestanding Stars and Stripes should be planted on the lunar surface – as a symbolic gesture, rather than one of sovereignty – with a stainless-steel plaque more representative of mankind as a whole: 'HERE MEN FROM THE PLANET EARTH FIRST SET FOOT UPON THE MOON, JULY 1969 A.D. WE CAME IN PEACE FOR ALL MANKIND.'

An engineering solution for erecting the flag was devised by Jack Kinzler, director of the Technical Services Division at what later became the Johnson Space Center. The clock was ticking fast to get a workable solution in place. First of all, the assembly had to be easy for the astronauts to operate in their cumbersome pressurized spacesuits and gloves. The lack of any atmosphere meant

---

* Rock is perhaps too grand a word, as they are rice-sized grains.

ASTRONAUT REACH CONSTRAINTS

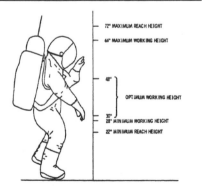

72" MAXIMUM REACH HEIGHT

66" MAXIMUM WORKING HEIGHT

48"
OPTIMUM WORKING HEIGHT

30"
28" MINIMUM WORKING HEIGHT

22" MINIMUM REACH HEIGHT

the flag would just limply hang down, so it would have to be engineered to look as though it were flying triumphantly. And every gram of weight and space on board had to be accounted for. It was decided to stow the flag assembly behind the side of the Lunar Module ladder, rolled up inside a metal case and insulating material to protect it from the intense heat of the rocket motors.

Previous page: **Apollo 17's Harrison Schmitt inspects a giant boulder**
Above: **Andy Warhol's** *Moonwalk* **(1987)**

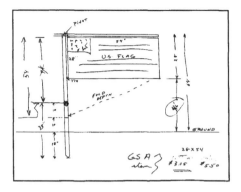

The flagpole assembly consisted of telescoping anodized aluminium tubes, with a steel spike at the bottom and a red ring painted on it, to give the astronauts a visual sense of how deep they had hammered it into the lunar surface. The nylon flag material was hemmed along the top, which Kinzler credited to watching his mother make curtains. A hinged telescopic crossbar was inserted with a stop catch to keep it horizontal at 90°. A sewn loop held the bottom of the flag to the pole. The entire flag assembly was rushed to the launch pad and installed on the Lunar Module only hours before lift-off.

It's estimated that the cost of the entire Apollo project was nearly $25 billion. The Apollo 11 flag and assembly came in at under $100: the flag costing $5.50 and the metal tubing around $75.

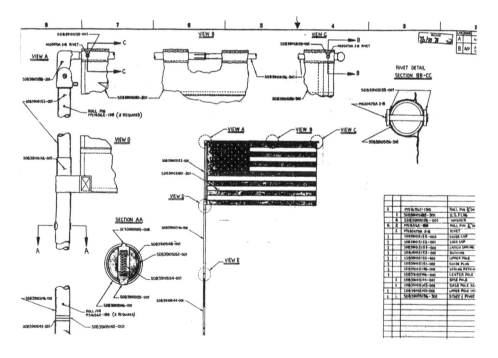

WHOSE FLAG IS IT ANYWAY?

'The Stars and Stripes to be deployed
on the Moon was purchased along
with several others made by different
manufacturers at stores in the area
around the Manned Spacecraft Center
near Houston. In order to attach the flag
properly to its aluminum staff it was
necessary to remove the binding and
labels. For this reason the name of the
manufacturer cannot be determined.'
NASA Press Release 69-83E, 3 July 1969

A mystery surrounds who made and
supplied the Apollo 11 flag. With such a
sense of national pride, multiple companies,
manufacturers and individuals have
staked a claim: Dolores Black, a seamstress
from Milwaukee working for the Eder Flag

Manufacturing Company, claims she was
solemnly asked to make the flag, saying
she even signed her name on the inner
seam to prove it. The obituary of Phyllis
J. San Antonio claims it was made by her
New York embroidery firm. NASA never
officially confirmed who manufactured the
flag, to prevent companies profiting from
their association. In the sleepy mill town of
Rhodhiss in North Carolina, the locals still
talk of how they produced the raw nylon
fabric for this most famous Stars and Stripes.
The town limits road sign even has a picture
of an astronaut with the flag which reads –
'U.S. Moon Flags Woven Here' – a claim to
fame proudly stitched into the town's history.
Their claim could well be right.

The details are muddled by two
conflicting accounts. Jack Kinzler confirmed
in an interview, and had scribbled on his

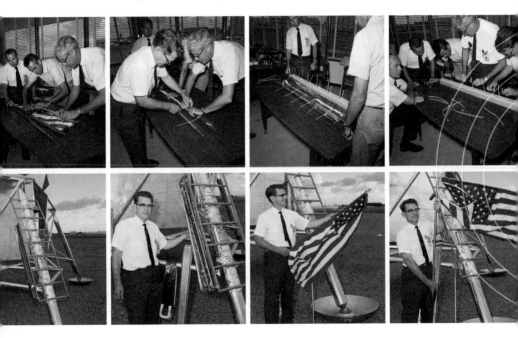

Above: **Jack Kinzler, David McCraw and team packing the Apollo 11 flag**

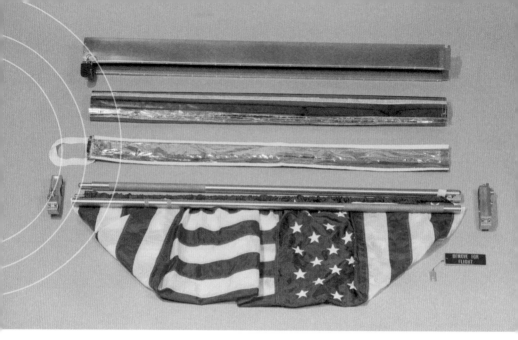

initial flag sketch, that the flag was purchased from a 'GSA' government supply catalogue for $5.50. The flag company Valley Forge claim to have been the supplier. Another story goes that three secretaries from the Johnson Space Center were dispatched on their lunch breaks to each purchase nylon flags of five foot by three foot from stores in the local Houston area. As it happens, the three secretaries all bought the flags from the same Sears department store, which at the time were supplied by Annin, another government flag supplier. In 1971, it was reported that 'Annin has produced all the flags carried to the moon by the Apollo astronauts'. Although still part of their publicized history, they are now slightly ambiguous in their claims, but have told me that they had a verbal confirmation at the time that an Annin flag was picked over a Valley Forge flag for the first historic moon mission. To make matters even more confusing, a nylon-textile-producing company called Glen Raven, who manufactured flag material for both Annin

and Valley Forge claim they wove the nylon fabric in Burnsville, North Carolina – not in the town Rhodhiss at all.

The details are now lost for ever in time and space. But the good news is that there are six flags on the moon, enough to satisfy everyone's claim without questioning it too closely.

What is the condition of the flags now? Some may still be standing. Some may have been blasted over by the Lunar Module rocket motors as they left the moon to return to earth. Have they been completely bleached white by solar radiation, as is commonly thought? For a space archaeologist it would be interesting to go back and see them again.

EPIC MEN OF FLESH AND BLOOD

The starting gun of the Apollo program was Kennedy's famous 'We choose to go to the moon...' speech at Rice University on 12 September 1962. Less well known is the speech written for President Nixon to read to the world in the event of disaster.

To   :   H. R. Haldeman

From: Bill Safire   July 18, 1969.

IN EVENT OF MOON DISASTER:

Fate has ordained that the men who went to the moon to

explore in peace will stay on the moon to rest in peace.

These brave men, Neil Armstrong and Edwin Aldrin, know

that there is no hope for their recovery.   But they also know that there

is hope for mankind in their sacrifice.

These two men are laying down their lives in mankind's

most noble goal: the search for truth and understanding.

They will be mourned by their families and friends; they

will be mourned by their nation; they will be mourned by the people of

the world; they will be mourned by a Mother Earth that dared send two

of her sons into the unknown.

In their exploration, they stirred the people of the world to

feel as one; in their sacrifice, they bind more tightly the brotherhood

of man.

In ancient days, men looked at stars and saw their heroes in

the constellations.   In modern times, we do much the same, but our heroes

are epic men of flesh and blood.

Others will follow, and surely find their way home.   Man's

search will not be denied.   But these men were the first, and they

will remain the foremost in our hearts.

For every human being who looks up at the moon in the

nights to come will know that there is some corner of another world

that is forever mankind.

PRIOR TO THE PRESIDENT'S STATEMENT:

The President should telephone each of the widows-to-be.

AFTER THE PRESIDENT'S STATEMENT, AT THE POINT WHEN NASA
ENDS COMMUNICATIONS WITH THE MEN:

A clergyman should adopt the same procedure as a burial at

sea, commending their souls to "the deepest of the deep, " concluding

with the Lord's Prayer.

'Yet across the gulf of space... intellects vast and cool and unsympathetic, regarded this earth with envious eyes, and slowly and surely drew their plans against us.'
H. G. Wells, *War of the Worlds*

MARTIANS WANTED

Ernest Shackleton famously placed a small advertisement for an Antarctic expedition crew in *The Times*: 'MEN WANTED for hazardous journey, small wages, bitter cold, long months of complete darkness, constant danger, safe return doubtful, honour and recognition in case of success.' Despite the dire warnings, hundreds of men eager for adventure and glory applied. Except they didn't. No one did. The truth is the advert never existed. We are easily fooled by this urban myth because we instinctively sense what such a call to action means and what's so seductive about the idea. We're emotionally drawn to the idea of 'the frontier', like a moth to the light, whatever the dangers may be.

Imagine a similar call to action for a journey to Mars sometime in the near future. What would be in store for you on a trip on the interplanetary liner HMS *Elon Musk*? A three year round trip, with one port of call. Permanent confinement to the stateroom during the traverse. Severe radiation risks from solar events spewing forth from our sun, and other cosmic nasties. Profound physiological issues – bone and muscle loss and macular degeneration caused by the ravages of a zero-gravity environment. Profound psychological discomfort would be a certainty – boredom, anxiety, cabin fever. No escape from your crewmates. No escape from

Above: **SpaceX's Interplanetary Transport System (artist's concept)**

yourself. No visual connection between you and the earth from the window, and limited communications back home to talk to friends and loved ones you're leaving behind.

And that's the good bit. When you arrive, life gets challenging. Mars's painfully thin carbon dioxide atmosphere would mean multiple problems to overcome, not least the difficulty of slowing a spacecraft down to reach the surface safely, a surface completely unable to sustain human life. This would mean near-permanent confinement inside some form of temporary pressurized inflated shelter, and what forays you would make outside would be in your spacesuit. Let's hope you've got a spare one. Maybe two. Hopefully some crucial infrastructure will be up and running for when you arrive: a vital source of power from a nuclear generator, perhaps. Oxygen and water would have to be reclaimed with equipment such as a MOXIE (Mars OXygen In Situ Resource Utilization Experiment), a machine designed at MIT soon to be tested on the new Mars 2020

Rover, that converts Mars's CO_2 atmosphere into oxygen. Frozen water ice would need to be mined from beneath the surface or extracted from the regolith.

The temperatures vary dramatically, depending on where and when you go. The equatorial regions can be as high as 20°C in the summer if you're very lucky, but on average it's going to be well below freezing and plummeting to -100°C or lower, colder than any of the research stations even in inland Antarctica. A lack of any sort of magnetic field means the surface is bombarded with intense solar and cosmic radiation. On earth, this is deflected by the magnetic field driven by our geological engine – our internal core of iron – a system that stalled on Mars sometime in its history. Any long-term Martian shelter would have to protect you. Are you going to go all that way to live underground? If you're unlucky, the thin winds, which you would hardly feel, will pick up the fine dust causing planetary dust storms, events that might shroud the view for weeks on end.

Above: **NASA Curiosity Rover's photograph of Gale Crater**

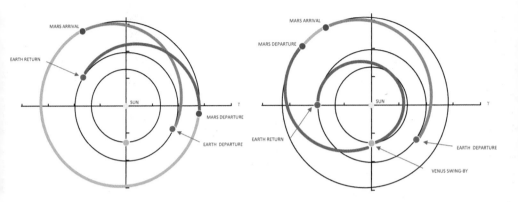

Above: **Two ways of getting to Mars: conjunction and opposition trajectories**

Once you're there, you'll have to stay for a while. Getting to and from Mars is entirely dependent on our orbits and proximities to each other, ranging from about 55 million km (opposition) to 400 million km (conjunction). There are small, strictly adhered to windows of opportunity for your departure and return – remember, you're aiming at a moving target from a moving target. You could choose an 'opposition-class' mission, staying 1–3 months, or 'conjunction-class', around 18 months, or extend your trip further with 'fast transit'. Or just stay forever.

Mars analogue missions here on earth always have a possibility of escape. An end in sight. A door back to the real world. Here on Mars, all doors are firmly bolted. If you're still alive and sane, your safe return may indeed be doubtful. You would in return be guaranteed some of the most spectacular views in the solar system. And 'honour and recognition' in case of success.

Do you still want to go? Of course you do. And the good news is they're already planning your trip.

A STRANGER IN A STRANGE LAND

Why do you want to go? Tourism? Exploration? Science? To fulfil mankind's destiny? To start a new life, never to return? For all the difficulties involved, and whatever your motivations, Mars is still by far the best option for a space destination beyond the moon. It's the only place that's close enough and where we can just about imagine our current or near-future technologies sustaining our frail bodies. The crushing hellish furnace of our other nearby neighbour Venus holds no respite for the weary traveller. Mars is familiar in many ways, a solid rocky surface you can stand on, albeit at 38 per cent of your weight due to its smaller size, which is useful for the rockets wanting to take you back home. There is evidence written into its surface of water, the single most important raw material needed to support us. The water no longer flows freely,* but waits to be mined, frozen beneath the surface. With water and CO_2 you can produce a whole host of useful things, including oxygen to breathe, and hydrogen and methane for rocket fuel. Without water, the dream is over.

---

* There is evidence of briny fluids flowing occasionally: for more information, look up 'recurring slope lineae'.

For the sightseer, Mars is rich in vistas, familiar to us from the millions of high resolution images sent back by the recent orbiters and rovers. There are the big must-see tourist destinations like Olympus Mons, the highest volcano in the solar system; the great Valles Marineris, a planetary gash extending about the width of the United States that puts our own Grand Canyon to shame; the polar ice caps of frozen CO_2; the craters, caves, lava tubes, dried up river valleys and dunes to explore. And then there are the alien landscapes, frozen in time on a stalled planet, that we have yet to imagine. On a clear day in winter, you would see the high, wispy cirrus clouds. And in the night sky, the Earth as a tiny pale-blue dot amid a billion pinpricks of light. Mars is a planet unspoilt.

For the scientist, Mars has become the ultimate multi-disciplinary destination, whether you're a meteorologist, geophysicist, engineer, chemist, geologist, astrobiologist or psychologist. Over the last sixty years we've sent dozens of spacecraft for a closer look – right now six orbiters are watching from the sky as well as two rovers still functioning on the surface, with more remote sensing missions to follow. Much has been revealed,

but much is still to be learned – information about the formation of the early solar system, for example, will still be visible in the Martian rocks; our own early history has been erased by the constant recycling of the dynamic Earth's crust. Central to all of the science, as well as our imaginings, is the enduring question of life. Did life propagate in a once warmer, wetter Mars? Is there evidence of that life now? Or are there conditions for life, of some sort, to still take hold? Humanity's most profound question will reposition our place in the universe once again. The parameters of this question are revealing themselves with a growing clarity.

For the cultural visitor, we have a long artistic relationship with Mars and its imagined inhabitants, through the great science fiction writers like Ray Bradbury, H.G. Wells and Robert Heinlein, as well as legions of TV producers and film-makers who have shaped Mars through the imagination. This 'other' Mars has always walked closely alongside its scientific twin. Recently Andy Weir's book (and film) *The Martian* revisited Daniel Defoe's *Robinson Crusoe* narrative, coming hot on the heels of the latest scientific revelation that Mars may, under certain conditions, have a form of running water – the salvation for any desert island castaway. The real and the imagined walk cheek by jowl. And sometimes they get confused: in the 1930s, it was reported that following Orson Welles's radio dramatization of *The War of the Worlds* there was mass panic across America as listeners thought the Martian invasion at Grover's Mill was real. You know the story I'm sure: people fleeing cities, having nervous breakdowns, a country in terror at the Martian invasion. It seems though that while there were a few dramatic headlines – 'Fake Radio War Stirs Terror Through U.S.' – it was the 'fake' news story of its day. There was no terror. There was no mass panic at all. Like Shackleton's fake advert, we are easily seduced by a glorious idea and dare not check too carefully for fear of breaking the spell.

Perhaps you're motivated to go by something grander. By wanting to bring about the dream of making humans a multi-planetary species – to create a new permanent home on our most extreme frontier, or as an insurance policy against some catastrophic disaster on Earth. At a conference in Guadalajara in September 2016, Elon Musk gave a deadpan, matter-of-fact presentation outlining his audacious plans for human colonization, proposing fleets of spaceships taking thousands of people to Mars, like the first wave of American settlers. Could this have been the PowerPoint presentation marking the beginning of a new chapter in human evolution? What future lies ahead for us after the first humans are born on another world? It could have been Tsiolkovsky standing there, or von Braun or Edward Nkoloso – all of whom had Mars in their sights, but who lacked the immense power of social media to capture the world's attention.

Whatever your reasons, Mars is now a realistic(ish) physical destination. More than just a stage to play out our fantasies. But it wasn't always that way.

A PALE RED DOT

The story of how we've unlocked Mars's secrets is as illuminating as the secrets themselves. For almost all of human history, all we've had to go on is a red pinprick of light in the sky on a dark, star-filled night. Mars wasn't a destination but part of a prevailing worldview – a celestial object on which civilizations could project the parochial beliefs of the time, the relics of such thinking still lurking in astrology today. The word planet comes from the Greek word meaning 'wanderer'. The Egyptian, Chinese, Babylonian, Greek and Roman civilizations were all aware of this enigmatic point of light with its unusual colour and its peculiar looping 'retrograde' path across the sky relative to the fixed stars. For an ancient astronomer, Mars's movement, in which it suddenly seemed to double back on itself, was a mystery. Even today it's not immediately obvious what is going on. Imagine athletes running around a running track. Mars is in an outside lane relative to us as we zip around the sun. As we overtake Mars on the inside, the observer here on Earth sees Mars appearing to move backwards in the sky, in a loop. It's one of nature's powerful illusions.

It wouldn't be the last time Mars would fool us.

ADJUSTING THE FOCUS

'All philosophy, said I, is founded on two things; an inquisitive mind, and defective sight...'
Bernard le Bouvier de Fontenelle,
Conversations on the Plurality of Worlds

Technology has brought Mars closer to us. The newly invented telescope was pointed towards Mars by Galileo Galilei in 1609 and gave us our first blurry close-up look. Mars was now a spot rather than a dot. Fifty years later, the Dutch scientist Christiaan Huygens brought Mars even nearer, noticing dark patches and a polar ice cap. A spot

earth (opposition). Schiaparelli knew the importance of a steady hand and clear mind when making astronomical observations. He banned himself from alcohol, narcotics and coffee. He described and drew dark striations that he saw on the Martian surface, which he called *canali*, simply meaning ditch, channel or trough.

But American astronomer Percival Lowell, working from his observatory high up in Flagstaff, Arizona at the turn of the twentieth century, conflated Schiaparelli's naturally occurring *canali* with the idea of engineered, intelligently designed 'canals'. Lowell didn't *think* there were canals on Mars, he *knew* there were canals. Schiaparelli's rather curved broad channels suddenly straightened out under Lowell's gaze, with the addition of round junction points (which Lowell termed 'Oases'). All this led to one indisputable conclusion: these canals were an irrigation system, part of a planetary geo-engineering project to protect and redirect water to Mars's arid regions by the Martian inhabitants. Of course, this was sensational news that was spread across the world. It's a salutary lesson: a clear desert atmosphere, a fine telescope and superb eyesight, which Lowell apparently had, are no defence against the distorting pattern-seeking power of the mind. Such bias is easy to see in others. Harder to recognize in ourselves.

was changing to a disc, with structure and features that changed with time and seasons, and of course with this came the speculation of life. With William Herschel (1738–1822) we entered a new era of observations, as he highlighted the similarities between Mars and the Earth. He said: 'I have often noticed occasional changes of partial bright spots…; and also once a darkish one… And these alterations we can hardly ascribe to any other cause than the variable disposition of clouds and vapours… And that planet has a considerable but moderate atmosphere, so that its inhabitants probably enjoy a situation in many respects similar to ours.'

LOST IN TRANSLATION

Over time, earthbound telescopes became increasingly powerful, but as well as bringing celestial objects closer they were limited by the distortion caused by light passing through our atmosphere, which in turn amplified the creative power of the human brain to fool itself. In 1877, the Italian astronomer Giovanni Schiaparelli inadvertently caused one of modern science's greatest misunderstandings, when observing Mars at its closest point to

Above: **Herschel's drawings of Mars**
Right: **Lowell's Martian canals**

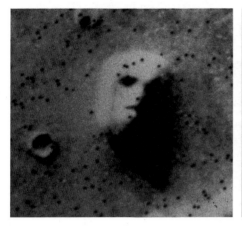

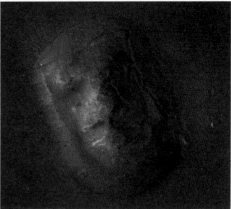

Such yearnings aren't confined to history. Richard Hoagland, an American conspiracy theorist, is still out there promoting the 'Face on Mars' phenomenon: a raised mesa in the Cydonia region, which when photographed by the Viking Orbiter in the 1970s resembled a Sphinx-like human face looking at the sky. Hoagland interpreted this as evidence of alien design. It's clearly a case of *pareidolia*, a well understood phenomenon whereby the brain sees faces in random visual noise where none exist. Even though the mesa has been re-photographed in superb clarity revealing nothing at all, the die-hards will continue seeing what they want to see. It's not just faces Hoagland's spotted – he's seen various architectural ruins, pyramids and evidence of the rubble of an entire ancient city. All pointing to evidence of a Martian civilization. It's entertaining stuff, and will give you plenty to think about when you get there.

Seeing is very often believing. But believing doesn't make things true. Occasionally bits of Mars come to us as meteorites, ejected from the surface of the planet by impact events. These have found their way to the Earth, across millions of miles of space, lying undisturbed only to be discovered by meteorite hunters. In the 1980s the Allan Hills (named after the region in Antarctica where it was found) 84001 meteorite was discovered. An interesting rock for a whole host of reasons, not least because of what was seen under the electron microscope over a decade later – the image of what looked like a minute worm-shaped structure. Could this be some sort of fossilized nano bacteria? The smoking gun, the answer to the great scientific question – evidence of past life on Mars? In 1996, Bill Clinton made a televised announcement about the findings from the south lawn of the White House. Sadly, it was generally agreed by the scientific community to be a resounding no.

Canals, swathes of vegetation and entire civilizations have featured heavily in the Mars story. The great American astronomer and writer Carl Sagan said it best during his Christmas lecture at the Royal Institution: 'Where we have strong emotions, we're liable to fool ourselves.'

Above: **The 'Face on Mars' is unmasked**

Left: **Meteorite ALH84001**

Wernher von Braun

Das Marsprojekt

Studie einer interplanetarischen Expedition

Ein Sonderheft der Zeitschrift „WELTRAUMFAHRT"

GETTING YOUR ASS TO MARS

After centuries of looking at Mars, going there is now a possibility. These days it seems anyone who's anyone has an extravagant plan to open up the new Martian frontier. Buzz Aldrin is rarely seen these days without his motivational *'Get Your Ass to Mars'* T-shirt. The Dutch entrepreneur Bas Lansdorp's 'Mars One' plan to send a colony of settlers on a one-way trip to the red planet began in 2012 with a call for colonists. There was no shortage of volunteers, with 200,000 people signing up for the trip. Some of these plans are short on detail and big on hyperbole; some are more realistic than others. Robert Zubrin, the founder of The Mars Society, which runs the Mars Desert Research Station in Utah, drew up a detailed low-cost plan called *Mars Direct* in the 1990s. Zubrin remains a passionate advocate of getting human to Mars, and is a vocal critic of NASA's own Mars plans, which he accuses of being wasteful, bureaucratic and lacking any sort of real focus.

But the grandfather of all Mars plans was Wernher von Braun's *Das Marsprojekt*. In 1952 he outlined, in considerable detail, the mathematics and engineering considerations of such a venture. Von Braun's idea was based on a flotilla of spacecraft with colonists moving en masse to Mars, rather than a single spacecraft. It was a bold visionary idea, ahead of its time but a foreshadow of a similar idea that the new rocket-building visionary of our age, Elon Musk, has recently run with.

1. SPACEX'S INTERPLANETARY TRANSPORT SYSTEM

'It'll be really fun to go.
You'll have a great time.'
Elon Musk

Everything Elon Musk has done up to now has been a lead-up to his ultimate goal of turning human beings into a multi-planetary species. The development of SpaceX's new rocket technology is central to this. In fifteen years, he's revolutionized the industry, bringing down the cost of getting into space with his reusable, self-landing rocket stages, as well as completely redefining the electric car, battery and solar panel industries in his spare time. All of this innovation is ultimately aiming to improve life on planet earth, as well as taking us out to the stars.

Musk's Interplanetary Transport System, which he outlined on stage in 2016, is an outrageous vision to transport large groups of people to Mars and beyond. It's as much an exploration of the economics of colonizing Mars as it is about the space architecture – how to eventually bring down the cost of a ticket to Mars (and back) and put it within the reach of mere mortals like you – in the hundreds of thousands of dollars rather than the millions. Here's how it works: a huge carbon-fibre booster will launch a spaceship designed to carry a hundred or more people into orbit. The booster will then return to the launch pad in the manner of the current Falcon 9 rocket and pick up a fuel tank, returning to orbit to fuel the spaceship. And then off you – and hundreds of other ships like it – go. The passenger ships are designed to be as pleasant as possible. There would be a vast cupola-like viewing window at the spacecraft's tip, behind which there would be a communal open space in which people float around, enjoying doing whatever people do on the way to Mars. Think a luxury liner, rather than a cramped airless capsule. Once on Mars, you won't need a separate booster to get you off the ground again – with the gravity well being only 38 per cent of Earth's, the spaceship itself has all the power you need – so long as there is fuel. As this is an interplanetary system, this isn't just about Mars – trips around the solar system, to the Jovian moon of Europa and Saturn's moon Enceladus, are all possible stop-off points too.

It's a wild futuristic vision, that carries with it a long tradition of wild futuristic visions that have never got beyond the fantasy stage. But Musk, as we know, is a doer as well as a dreamer. So however pie-in-the-sky you think all this might be, he's a difficult man to bet against.

2. NASA'S JOURNEY TO MARS

NASA's wants to send astronauts to Mars orbit by the 2030s. Will that happen? Will they be pipped to the post by SpaceX or Blue Origin? Is the political will there? As ever, much of NASA's direction of travel will be at the whim of the changing American political cycle. What the world will look like next week, let alone in a decade, is anyone's guess. But at least two crucial parts of the architecture are being built: the SLS heavy-lift rocket and the Orion spacecraft. The first big test of the new Orion spacecraft to the moon (Exploration Mission-1) is slated for 2019. But a trip to Mars and back will require orders of magnitude more work and more sophisticated space hardware development than that. Orion isn't the vehicle that's going to get you onto the Martian surface. In the meantime, NASA's *Journey To Mars* portmanteau is shifting its emphasis towards NASA's *Deep Space Gateway* – using the space between the Earth and the Moon as a proving ground for building and testing new technologies with the goal of a crewed spaceport in lunar orbit. In the meantime, scientific work will continue in low earth orbit on the ISS until as late as 2028.

NASA's *Journey to Mars* might not have stalled. It's just that the route NASA plans to take today will be subject to further changes and diversions. Plan accordingly. As Robert Zubrin points out, how soon we get there boils down to us: 'Where there's a will, there's a way. Where there's not a will, there's no way…'

How badly do we want to go? That's a decision we have to make.

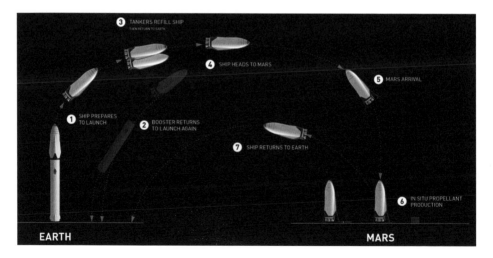

Above: **SpaceX's Mars ITS architecture**

Above: **A 'pale blue dot': the view of earth from your new home on Mars**

10 EXCITING PLACES TO VISIT IN THE UNIVERSE

'All these worlds are yours – except Europa. Attempt no landings there…'
Arthur C. Clarke, 2010: *Odyssey Two*

The first planet beyond our solar system was discovered in 1992. Since then we've found around 4000 more candidates, and counting. As I write this, the discovery of the 'Trappist' suite of exoplanets has been announced, and closer to home the Cassini spacecraft has confirmed that Saturn's moon Enceladus has – along with water and energy – the right chemistry for life. The universe is more fascinating than we can ever imagine.

Astrobiologist and planetary geologist Dr Louisa Preston of Birkbeck, University of London, gives us her 'top ten places you'll never visit' out of the countless billions of destinations out there waiting to be discovered.

JOURNEY TIME CALCULATOR
'New Horizons' is currently our fastest-moving spacecraft, travelling at 36,373 mph, and has been used to calculate all travel times. Various exotic methods of propulsion like magnetohydrodynamic drives are still some way off.

1 ENCELADUS: Over a hundred geysers of ice and gas are erupting from the southern pole of this 500-km-wide moon of Saturn, coming together to form a colossal plume spewing over 400 km into space – about the distance from London to Newcastle. The source of these geysers lies beneath the icy crust – a vast ocean full of the chemical ingredients for life, and possibly life itself. Take your warmest clothes: the average temperature is a chilly -198°C, and snow falls across the surface from the plume.
Journey time: 2.5 years

2 THE SUN: The largest body in the Solar System is 93 million miles from the earth and a source of heat and light for life as we know it. You could peek beneath the surface of this ordinary G-type star and watch the creation of the very elements that make-up the universe. Pack strong clothing – enough to protect against heat that can melt iron. Bring sunscreen.
Journey time: 106 days

3 EUROPA: What secrets are hiding beneath the thick icy shell of Jupiter's moon? A salty ocean, two to three times the volume of the Earth's, may support extreme-loving organisms or vast ecosystems around possible black smoker hydrothermal vents on its floor. The only way to know for sure would be to drill through the ice, take a submersible and explore the hidden depths.
Journey time: not recommended (see above)

4 TITAN: Saturn's largest moon is nearly 900 million miles away – weirdly familiar

yet completely alien, protected by a thick atmosphere and covered in mountains, valleys and bodies of liquids, but using a different chemistry. It has all the basic elements necessary for life but instead of water, the lakes are filled with liquid ethane and methane – what would it be like to sail across these? The best thing, however, is that thanks to the thick atmosphere and low gravity we could strap on wings and fly.
Journey time: 2.4 years

5 PLUTO: Only recently revealed by the New Horizons space probe which crossed 3 billion miles of space to show us what it looked like, a remarkably exciting reddish-brown world with flowing nitrogen ice, frozen water, mountain ranges, tar-like rain and an exotic carbon-rich surface chemistry. Even in the farthest reaches of the Solar System this frozen world has habitable environments, so who knows what else we will discover on this once-upon-a-time planet.
Journey time: 9.5 years

6 PROXIMA CENTAURI: Our nearest star has a roughly earth-sized planet orbiting in its habitable zone, called Proxima B. This exoplanet circles its red dwarf star far more closely than Mercury orbits our Sun, but could be at just the right distance for liquid water to exist on its surface. It is a mere four light years away. But if this world was rocky, had liquid water and an atmosphere, who knows what, or who, might be able to exist on it. Maybe you.
Journey time: 73,796 years

7 KEPLER-452B: The earth's older, bigger cousin – an almost earth-sized exoplanet in the habitable zone of its sun-like star, 1400 light years away. It is 1.5 billion years older than our planet and the sun. This world could be extremely exciting as life, should it have arisen there, has had a longer time to evolve, providing a glimpse into our possible future.
Journey time: 26 million years

8 THE EAGLE NEBULA: A 5.5-million-year-old cloud of molecular hydrogen gas and dust, 7000 light years away. To soar through this region would be to marvel at stellar nurseries such as the Pillars of Creation, to witness stars being born and to dodge the dying gasps of older stars as they go supernova, ejecting radiation and powerful shock waves.
Journey time: 130 million years

9 THE MILKY WAY: Every image we have of our galaxy isn't real, as we have never left it. For us, snuggled inside the Orion arm of the Milky Way it is like trying to take a photo of the outside of our house while standing in the kitchen. Our Solar System is about 24,000 light years from the outer edge, so it could take up to 443 million years for us to reach it – but it would be a spectacular sight to float above the plane of the Milky Way and take the first-ever photo of the billions of stars, gas and dust we call home.
Journey time: 443 million years

10 SAGITTARIUS A*: This is a supermassive black hole in the heart of the Milky Way galaxy, a dark point 15 million kilometres wide and 5 million times the mass of the sun where nothing, not even light, survives. Approaching the event horizon, the light of the universe distorts and is squeezed into a thin band, before vanishing completely as you head towards the crushing central singularity. Here, powerful tidal forces stretch you into atomic spaghetti, eventually ripping you apart. At 26,000 light years from the earth, it would take a while to get there, but time is relative. And it would be a spectacular way to die.
Journey time: 480 million years

ENVOYS: EXTENDING OURSELVES

'The only true voyage of discovery, the only fountain of Eternal Youth, would be not to visit strange lands but to possess other eyes, to behold the universe through the eyes of another, of a hundred others, to behold the hundred universes that each of them beholds, that each of them is...'

Marcel Proust, À la recherche du temps perdu

While a human being has not yet ventured beyond the moon, legions of our robotic spacecraft have scouted ahead, extending our senses across the solar system. All the planets have now been visited, transforming our knowledge and overturning our expectations. Lately it is these envoys that have been grabbing all the headlines – Rosetta and its lander module Philae have explored a comet; the Cassini probe has revolutionized our understanding of Saturn, its rings and its moons, and ended its life crashing into the planet. The New Horizons' spacecraft photographed Pluto in exquisite detail for the first time in 2015.

Such spacecraft have stretched our engineering capabilities to the limit. Building a vessel to travel even the relatively short distances to the nearest planets is hard. It was especially hard in 1964, when no one had built a successful interplanetary spacecraft.

In the first few years of the 1960s, the Soviets had already launched a number of robot emissaries to Mars. All of them had failed, but they hadn't given up. America in the meantime was basking in the success of Mariner 2 (Mariner 1 was destroyed after a launch anomaly), the first spacecraft to ever reach another planet, which in 1962 successfully journeyed past Venus, sending back data about the planet, before being cut adrift, locked in an endless orbit around the sun. Mariner 2, like a licked finger held up to the solar wind, had given American scientists a flavour of what was possible. Mars was now in their sights. And the building of such a craft, a ship that could sail further than ever, lay with the shipbuilders of NASA's Jet Propulsion Laboratory (JPL), nestled in the foothills outside Pasadena and led by William H. Pickering. But this project would require a complete redesign. Something more sturdy and watertight, for a journey significantly more treacherous and complex. And with a lot more at stake.

MARINER 4

In 1964, our best images of Mars were still fuzzy blobs of light and dark. What was needed was a camera that could sail right up to Mars itself, take some snaps and report back. Film of course would be of no use here. The camera and the spacecraft were never coming back, and so what became the world's first digital camera was conceived and built by a team at JPL, led by the esteemed Caltech physicist Robert Leighton. It would be a camera that could turn an image into digital data recorded on a magnetic tape, which could then be beamed back home. Twenty-two 200 by 200 pixel photographs would be taken of the planet's surface during a single fly-by, to give us ten times better resolution than we could see from earth. A hundred metres of magnetic tape was used to capture the images, averaging around $3.8 million a picture. There would be no second chances. No forgetting to take the lens cap off.

Work was started on a pair of identical twins – Mariner 3 and 4. The core of these two spacecraft was an octagonal magnesium 'bus' frame which housed the electronics, and hydrazine course-correcting propulsion systems – this was the architecture that became the signature blueprint design of JPL's spacecraft that were to follow. Four extra-large solar panels jutted out, giving it the appearance of a giant flying windmill.

As well as the space camera system, there was a host of other scientific equipment: a magnetometer, and cosmic ray, dust and radiation detectors, ready to test the waters and take the pulse of Mars. Perhaps the most important innovation was the navigation system. Like mariners here on earth, the Mariners would be guided by the stars. In this case, the bright star 'Canopus' would be the guiding light.

Launch of Mariner 3 on an Atlas rocket was on 5 November 1964 at Cape Kennedy, but the protective 'shroud' on top of the booster failed to jettison as it left the atmosphere, and it was lost for ever. The mission's success was left to the twin Mariner 4. The shroud's technical problems had to be solved in record time, in order to launch it during the short window of time available to make it to Mars before it sailed out of reach. Three weeks later, and with the Soviet Zond Mars probe scheduled for launch only days away, on 28 November Mariner 4 was off on its 228 day journey to Mars, with 325 million miles of deep space to navigate. It managed the journey with only one mid-course correction en route.

One of the main concerns was control of the digital tape recorder. It would sweep across the planet, snapping the series of images of about 1 per cent of the Martian surface. With its primitive computer, sending commands to the spacecraft was a perilous procedure. One major worry was that a command to stop the tape recorder from running (the 'off button') wouldn't work, and that the images would be wiped over, or not leave enough room for all the pictures intended. Mariner 4 flew by Mars on 14 and 15 July within a range of between 9846 and 12,000 km, taking twenty-one (and a few lines of number twenty-two) pictures in a twenty-five minute session. After reappearing from behind the planet, it began the arduous task of sending back the data collected at 8 1/3 bits per second: it would take many hours for the first image to be downloaded and processed, and nearly three weeks to transmit all the data in duplicate. For the first time in human history our eyes had been on another planet, and the world was waiting impatiently for the results.

PAINTING BY NUMBERS

Expectation was high from the world's media, anxious to see the first picture of Mars from space. But the stream of data that chugged back to JPL via Goldstone tracking antenna in the Mojave Desert as a series of ones and zeros would still have to be processed by computer to create the actual images. It would be days before all the images could be seen. Eager to see if the tape recorder had worked as planned, engineer Richard Grumm and John Casani, who had been working on the tape recorder, decided to jump the gun and get creative. Data from Mariner's camera was being converted into printed numbers on paper and spat out by a machine. The paper for one image was cut into long strips and arranged vertically side by side on a large room dividing board – the team began to colour in the strips using pastel crayons hastily bought from a local art shop. It was the most expensive painting-by-numbers art project ever created.

As the engineers began hastily colouring away, the Martian landscape swam into view – at the bottom, the numbers that denoted the blackness of space gave way above to the red and brown and yellow hues of the planet's limb (edge) and surface. The hand-shaded picture turned out to be remarkably accurate when compared with the actual photograph processed some time later.

For the engineers involved, this was a way of demonstrating the tape recorder had worked. But the significance of what was unfolding as the image took shape drew crowds of JPL workers to gather round and watch, while an armed guard prevented the eyes of the press seeing what was to become our first (beautifully false) colour view of another planet taken from space.

The picture is now framed outside what was the office of William H. Pickering, the founding father of JPL. Our first image of Mars from space was realized not by a machine, but by the flow of human hands. Its place in space history is clear. But it is also the most important piece of American modern landscape art ever created.

Above and opposite: **Hand-colouring the Mariner 4 data**
Right: **Mariner 4 photograph of Mars**

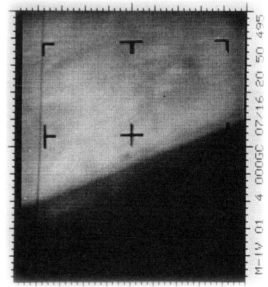

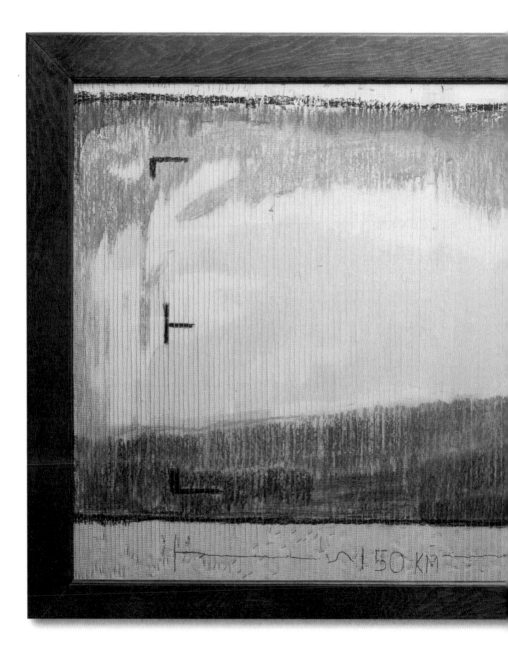

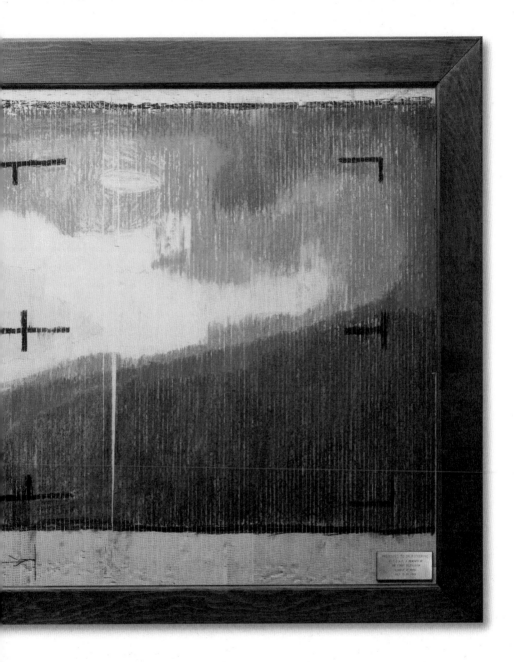

MESSAGES IN BOTTLES

While our robotic spacecraft's primary function is to explore on our behalf, it can also carry with it something of us. The Zond 2 Soviet mission to Mars supposedly carried a commemorative pennant destined for the Martian surface. Closer to home, the LAGEOS (Laser Geodynamics Satellite) 'Space Geodesy' was launched in 1976, a strange-looking disco-ball built at the Marshall Space Flight Center designed as a laser-ranging reflector in orbit, bouncing back laser beams to make very precise continental drift measurements on earth. It has a solid aluminium shell covered in reflectors and a heart made of a solid cylinder of brass. On board is a small plaque designed by Carl Sagan – a time capsule for the people of earth in the year 8 million, which was how long it was thought LAGEOS would remain in orbit. The plaque has a map of the movement of the continents over time and the numbers 1 to 10 in binary. The NASA press release ends with: 'Whoever is inhabiting earth in that distant epoch may appreciate a little greeting card from the remote past.'

Pioneer 10 and 11, launched in 1972 and 1973, were our first interstellar messages in bottles – thrown out into the deep water, dispatched to photograph and explore Jupiter and the interstellar medium, and now lost for ever. On board both spacecraft was the famous gold-plated plaque designed once again by Sagan and astronomer and SETI (Search for Extraterrestrial Intelligence) founder Frank Drake, which among other references to our planet had a line drawing of a naked man (waving) and woman drawn by Carl Sagan's then wife Linda Salzman. What was deemed suitable for any alien civilization was considered too risqué for some of the inhabitants of planet earth, and genitalia were swiftly airbrushed out in the American press; there was also criticism by those who saw the image as portraying the woman to be subservient to the man.

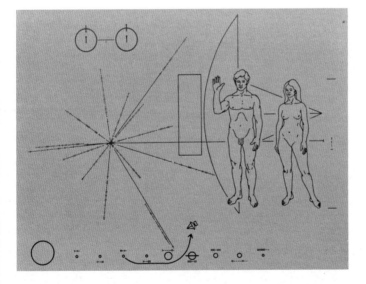

Opposite: **LAGEOS**
Above: **The Pioneer plaque**

THE FURTHEST WE'VE EVER SAILED

'I must go down to the seas again, to the
   lonely sea and the sky,
And all I ask is a tall ship and a star to
   steer her by,
And the wheel's kick and the wind's song
   and the white sail's shaking,
And a grey mist on the sea's face, and a
   grey dawn breaking.'
John Masefield, *Sea Fever*

The two Mariner-class spacecraft named
Voyager 1 and 2, launched in 1977, have set
the benchmark for deep-space exploration.
What was originally intended as a grand
tour of the outer planets of Jupiter, Saturn,
Uranus and Neptune has became a symbolic
extension of our most profound quest to
explore. It's a quest that is still going forty
years later. If you go into the Theodore von
Kármán lecture theatre at JPL in Pasadena, on
the left-hand side wall you can see an insect-
like Voyager spacecraft up close. It's built from
flight spares, so who knows, maybe if you
asked really nicely you could get them to fire

it up for you. It's dominated by the huge white high-gain antenna dish sitting on a decahedral bus similar to the Mariner, which contains the brains of the spacecraft. Jutting out of this body like spindly limbs are the 13-metre magnetometer boom, the plutonium-powered RTG (radioisotope thermoelectric generator) battery kept well away from the main body of the craft and, on another boom, the various scientific instruments including the scan platform with Voyager's eyes and ears. It's a remarkable piece of engineering, designed to function unsupported in the most hostile environment of deep space.

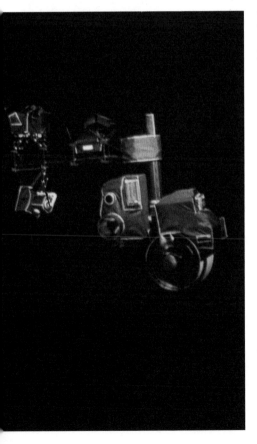

But how can you travel such vast distances using only rocket propulsion? How could you carry enough fuel and escape the gravitational pull not just of the earth but of the sun? At the time, getting beyond Mars and Venus seemed impossible. The solution was 'gravity assist', which changed the way we think about travelling long distances and is important to know about if you want to travel to the outer solar system and beyond.

Let's assume you're in your spacecraft. As you approach a planet (which itself is in motion, orbiting the sun at tens of thousands of kilometres per hour), as long as you're careful not to crash into it, you can use its gravitational pull, capturing some of that energy to 'sling-shot' you further and faster into space – perhaps towards another planet, which in turn can bounce you out even further. It's a useful trick that every space explorer should know and one that requires no fuel. In 1961 a young maths PhD student called Michael Minovitch who was studying at UCLA figured out the fiendishly difficult 'three body problem' in mathematics: in our case, how something small like a spacecraft is affected by the gravity of two more massive objects such as the sun and a planet. Minovitch calculated hundreds of different pathways to the planets, theoretically bouncing limitlessly from one to the other. In 1965 Gary Flandro, another young student, spotted one very special planetary alignment of Jupiter, Saturn, Uranus and Neptune, all of which were nicely lined up in a specific time period between 1975 and 1978. With the maths in place and the planets lining up in a window of opportunity that wouldn't open again for another 175 years, the engineers at JPL would only have one chance. A golden opportunity to sail further than we'd ever been before. The engineering challenges of such an ambitious mission were daunting and were overseen by project manager and engineer John Casani who had worked on the Mariner programme.

HELLO FROM THE CHILDREN OF PLANET EARTH

'I get a gasp of surprise at two numbers
that I give out when I describe the project.
The first is: I worked on something that's
going to last for a thousand million
years... But we had to do it in six weeks –
and that's when the real gasp comes.'
Artist Jon Lomberg, one of the instigators of
the Voyager spacecraft's Golden Records

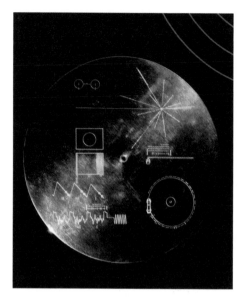

Here was a chance that two man-made
objects would be ejected from our solar
system and essentially last for ever,
orbiting the centre of our Milky Way galaxy
with the tantalizing prospect of one day
being intercepted by another space-faring
civilization. John Casani approached
Carl Sagan to provide something like the
Pioneer plaque that could be carried on
the Voyagers – Carl worked with a select
group of scientists, academics and thinkers
to come up with an idea for something. In
January 1977, the astronomer Frank Drake
had the idea of creating a phonograph record,
which coincidentally was celebrating its one
hundredth anniversary since its invention by
Thomas Edison in 1877. The physical groove
of a record is the perfect medium to store
information indefinitely, something that can't
deteriorate or be wiped over like a magnetic
tape. It would be made from copper and
gold-plated. You might raise an eyebrow at
the 1970s technology, but as we all know vinyl
never goes out of fashion.

But what to put on the ultimate mix
tape? One hundred and twenty-two images
reflecting the diversity of life on earth
were encoded in the audio spectrum – a
breastfeeding mother; an eagle in flight;
the diverse landscapes on earth – as well as
greetings in fifty-five languages, including:
'Paz e felicidade a todos' spoken by a young
Janet Sternberg. On the cover also
coded clues to the origins of this unusual
spacecraft. And music. Music seemed like the
perfect universal language, common to all

civilizations here on earth, but with limited
space and only six weeks to do it, how do
you choose the music that best represents
all cultures? Consulting various experts from
around the world, music was chosen to reflect
every continent and every epoch, everything
from Beethoven, to Bach, to Chuck Berry, to
Azerbaijani bagpipe music, a process which
became a frantic exercise in ethnomusicology,
speed and politics. The Beatles, for example,
couldn't be used because the request for
'Here Comes the Sun' was turned down by
the band's record company – presumably they
couldn't get past the legal ramifications of 'in
perpetuity, across the known Universe'. Ann
Druyan, the creative force behind the project,
who would later marry Carl Sagan, reflected
on the final piece – the Cavatina movement
from Beethoven's String Quartet No.13 – in
a BBC radio interview. She had been looking
through excerpts from Beethoven's diary: '
"Will they like my music on Venus? What will
they think of it on Uranus?" At last, a way to
respond to that impulse. To that question that
Beethoven felt so very long ago.'

Cavatina.
Adagio molto espressivo.
sotto voce

sotto voce

sotto voce

sotto voce

For the first time, these two spacecraft were to reveal the realities of the distant outer planets. The swirling atmosphere and famous red spot on Jupiter – a raging storm – were seen in sharp clarity. Jupiter's moons, first seen as dots by Galileo Galilei centuries before, were now visible up close, with Io's huge volcanic plume, 270 km high, seen for the first time. Saturn's rings could be seen for what they were – as multiple rings within rings, with tiny 'shepherd' moons embedded within them.

Here, the two Voyagers parted company. Voyager 1 diverted to investigate Saturn's moon Titan, heading up out of the plane of the solar system, its last official planetary encounter, while Voyager 2 carried on to explore Uranus, and then Neptune – the giant blue planet guarding the edge of the solar system like a sentinel, with its moon Triton with frozen nitrogen ice caps and nitrogen geysers. The Voyagers completely rewrote how we saw the outer solar system. Not just as distant static objects but as destinations. The solar system was alive.

Some twelve years after this epic journey started, the final picture from both spacecraft was taken by Voyager 1 on Valentine's Day 1990. In its elevated position, its cameras were turned around and a group family portrait was taken of all the planets together, a mosaic, all except Mars and Mercury, which were obscured by the brightness of the sun. It was the picture of earth that grabbed the world's attention. A tiny blue dot suspended in a beam of scattered light, the earth as a mere pixel seen from over 4 billion miles away. It is one of the most important photographs ever taken. The ultimate exercise in perspective. An image that shows the stark reality of our place in the cosmos, set free from our 'imagined self-importance'.

Forty years on, the two Voyager spacecraft are still going. Their instruments are gradually shutting down. The engineers who built and cared for them can mark the milestones in their lives against the Voyagers like a yardstick – children born, falling in love, marriages ended, friends and relatives passing away. Right now the Voyagers have passed the 'heliopause' – that demarcation line like a lapping shore, where the solar winds can no longer be felt. They have crossed a point that marks the shallow waters of our solar system into the deep waters of the interstellar medium – the space between the stars. Voyager 1 is now 12.9 billion miles away and Voyager 2 is 10.6 billion miles.

Out in the benign depths of the Milky Way, they will not rust or decay. These two spacecraft with their 'murmurs of earth' will outlive everything we know. Our most distant emissaries travelling for ever through an ocean of space and time, a snapshot of our moment in the sun. The Voyagers weren't just built to last – unlike us, they were built to last for ever.

'I can sense stars, and their whispers amid the roaring of our own Sun.'
Voyager 2, via Dr Paul Filmer

Opposite: **Earth photographed as a 'pale blue dot':**
Top: **By Voyager 1** Bottom: **Through Saturn's rings by the Cassini spacecraft**

TRAVEL GUIDE: HEAD OF SPACE AGENCY

Name:
**Professor Johann-
Dietrich 'Jan' Woerner**
Profession:
Civil Engineer
Claim to fame:
**Director General of the
European Space Agency**

What can I do for you?

**There are few better people
for some final thoughts
on leaving the planet than
the Director General of the
European Space Agency. I
read somewhere that you
used to build rockets as a
kid?**
Yes. A lot. Different types
with different propellants – I
tried the liquid propellant, I
did not succeed, but a lot of
different solid propellants.

**Were you aware of the
history of spaceflight?
Particularly early German
rocketry?**
Of course. Whether Germans
or non-Germans, this was all
close to my heart since the
early 1960s. Space for me
started in 1957. I was three
years old and my father took

me on his arm and said '*Okay,
look, up there...There's Sputnik!*'
and he said it so strongly I
really believed I could see it.
So that was the beginning.
And afterwards I was really
thrilled by space, so I built
rockets and all of this stuff.

**Those early rocket pioneers
were inspired particularly by
the science fiction of the day.**
Jules Verne's *Journey to the
Moon* was also one. And
interesting enough, the
launch site was in Florida! All
of these different visionaries
inspired me.

**As a kid it seems you were
already in space.**
Yes. We are all in space.
We are astronauts on the
spaceship Earth.

**I think people wanting
to leave the planet forget
they're already in space.**
Exactly.

**What is it for you that
makes a good astronaut?
There is a common thread,
a confidence and good
humour running through
the handful of astronauts
I've met.**
It's a balance between self-
confidence and team ability.
So he or she must be self-
confident and strong enough
to take responsibilities, but
able to work together with
others. And this balance, I
think this is something that
makes a good astronaut,
because we need people who

can make decisions while
they're on a trip and at the
same time they should be
able also to follow others
and this is a challenge for
people. The other things you
can learn how to deal with,
scientific things, but to have
a balance between these two
behaviours, this is something I
think you cannot learn. Either
you have it or you don't.

**We've got the ISS now until
hopefully 2028. We hear a
lot from NASA, Elon Musk
and others about going to
Mars. I wondered if you
could give us your thoughts
on where we're heading
for. 'Space 4.0' – the next
chapter.**
It is very clear to me that you
must learn to travel through
space. We'll travel beyond
the moon. We'll travel even
beyond Mars. It's only a
question of timing. I'm quite
sure that humans will do
that. If we survive. Stephen
Hawking said it was the other
way round, '*In order for us to
survive, we have to leave*'. I say,
in order to leave the earth, we
first have to survive.
    Right now, I think going
to Mars is very difficult. It's
much more difficult than to
go to the moon. You can say,
okay, it was difficult to go to
moon some fifty years ago,
but the difference is that we
know the dangers of going
to Mars from a health point
of view. Also, if you go to the
surface of Mars you need a
rocket to come back. I cannot

imagine that the journey will be possible with humans in the next ten to fifteen years. The Americans now have changed their wording, they are not talking about a 'Journey to Mars', they now say 'Deep Space Gateway'. I think one can be a visionary, but this is really outside of any vision.

Therefore, I think it makes sense to go to the moon. In any case we have to develop technologies and the moon is a perfect stepping stone. It's also interesting from a science point of view, developing technology and for space tourism maybe. So I think this is vision enough: let's go to the moon in the next ten to fifteen years, and then we can also have dreams about going further into the universe, but for that we need totally different technology and this is not on the table.

Humans will go. For sure. They will go to an asteroid, they will go to Mars, they will go to other places in the universe. I don't think that Mars is the ultimate goal – humans will go further.

And do the Americans agree? How closely do you work with NASA?
Very closely. They are also saying to go to Mars is a very long journey. They are talking now about the middle of the 2030s. I think this is the minimum time we need. We'll see what Elon Musk is doing, but all the spacecraft we have so far are not really a solution to go to Mars. We know that the Apollo astronauts were very lucky. If you go with an Apollo spacecraft to the moon and you get hit by a solar flare, then you are really burnt. So they were very lucky that they returned safely. But we should not take such a risk when we go for two years on a trip to Mars and back.

Is our greatest challenge in getting to Mars an engineering challenge? Or is it biological?
I think both. But I think the engineering we can solve by funding. If we do not have a totally different propulsion system, and are dealing with today's technology, then we still have the two-year timeframe. So I think the engineering and medical aspects are linked strongly with each other. If we can make the trip to Mars in a much shorter time, then of course the medical issue is reduced. But as long as we do not have that, then the medical issue is a very big one.

Presumably it comes down to politics as well. Who knows what the world is going to look like in twenty years' time? How frustrated are you by the short-sighted political cycles you have to navigate?
Well, political cycles and space do not fit very well together. But look at me, I am elected for four years, and if I say 'I'm only interested in things which can be done either in my first or second term,' then it would be ridiculous. So we need people who are thinking to prepare for the future in the long term.

There is a common yearning among space pioneers throughout history, and I think even yourself, who see human spaceflight as our destiny. Would you agree that dream is still alive?
I think it's still alive, I really believe it's still there. We have these enthusiasts, we have people with visionary ideas, we have people thinking far, far beyond their own possibilities time-wise and commitments-wise, and we need those people as dreamers from the future.

I wanted to give you the last word in the book. For those people who want to leave the planet, what advice would you give them?
Dream. And do it.

DESPERATE MEASURES

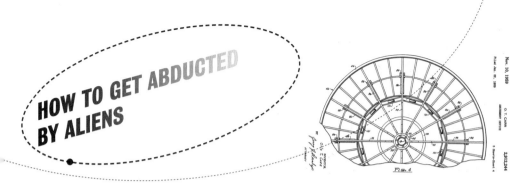

HOW TO GET ABDUCTED BY ALIENS

Douglas Adams in his five-point plan on how to leave the planet* in *The Hitchhiker's Guide to the Galaxy* was to flag down a passing flying saucer. Easier said than done, unless you're one of the many people who claim to have done just that.

It was late on 19 September 1961 when the aliens took Betty and Barney Hill. Driving home along a lonely road in New Hampshire, their journey was interrupted by a strange hovering illuminated disc. Barney could see a group of humanoid figures wearing sinister black uniforms watching them from inside. They sped off in the car, but retained no clear memory of what happened next – only through regression hypnosis in the months that followed did the bizarre story unfold: a rabbit hole of dreams, missing time and invasive medical procedures that was to firmly entrench the alien abduction myth into popular culture…

'I WANT TO BELIEVE'

The modern 'Flying Saucer' was born on 24 June 1947, when pilot Kenneth Arnold reported seeing a string of objects flying 'like saucers skipping over water'. He didn't consider the possibility of them being alien in origin, but thought at first they could be a flock of geese. From the moment it was reported, an entire science fiction and conspiracy subculture was born that captured the post-war paranoia of the day and piqued the interests of the military and of luminaries such as psychoanalyst Carl Jung. *Believing* in life beyond the earth – especially simple life – is easy. Most of us, without direct evidence, can happily assume the earth is not the only cradle of life in the universe. Believing in technologically advanced spacefaring extraterrestrial civilizations that have mastered travelling vast distances to whisk you away from the earth for some rectal-probing, clandestine agenda is harder. But not impossible – according to a recent National Geographic poll, a startling 36 per cent of Americans believe in UFOs.

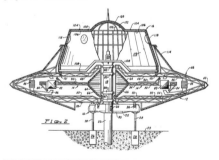

---

* By which this entire book was inspired.

SEAL · OF · THE · STATE · OF · NEW HAMPSHIRE ·
1776

BETTY AND BARNEY HILL INCIDENT

·

On the night of September 19-20, 1961, Portsmouth, NH couple Betty and Barney Hill experienced a close encounter with an unidentified flying object and two hours of "lost" time while driving south on Rte 3 near Lincoln. They filed an official Air Force Project Blue Book report of a brightly-lit cigar-shaped craft the next day, but were not public with their story until it was leaked in the Boston Traveler in 1965. This was the first widely-reported UFO abduction report in the United States.

2011

HERE I AM

When trying to flag down a passing flying saucer (as anyone who's hitch-hiked will tell you) it's good to be conspicuous. Having a sign helps. We are already doing a good job at advertising ourselves to anyone watching us – from our geography, to our atmospheric chemical signatures, to our radio broadcasts. Whether aliens will be tuned in to these, or indeed care, is an unknown. We could keep sending greetings, such as the Arecibo telescope broadcast sent towards a nearby star cluster in 1974, and the mix-tapes attached to our spacecraft. SETI has been listening for signs of intelligent life since astronomer Frank Drake's radio telescope search of the Tau Ceti and Epsilon Eridani star systems in 1960. The Russian entrepreneur Yuri Milner's $100 million 'Breakthrough Initiatives' programme is now investing heavily in a multi-disciplinary scientific search for life beyond earth. We'd all have to ignore Stephen Hawking, who has warned us of the dangers of advertising ourselves too freely – the history of first encounters between earthbound civilizations has often been disastrous. The fact that *they* have come to *us* means by definition they are more advanced than we are. Our 'imagined self-importance' maybe our undoing. Proceed with caution.

WHERE IS EVERYBODY?

How common are advanced civilizations in our galaxy? The famous Drake Equation, while not giving you an answer, does at least provide a sensible framework for asking the question and a starting point for scientists to argue over. It's a series of assumed important 'guesstimates' based on seven variables, which when multiplied together will spit out the number of civilizations in the galaxy, 'N'. We do know that the value of 'N' is 1 or more because here we are. Drake's conservative estimate is 50,000.

$$N = R_* \cdot f_\mathrm{p} \cdot n_\mathrm{e} \cdot f_\mathrm{l} \cdot f_\mathrm{i} \cdot f_\mathrm{c} \cdot L$$

$N =$ the number of civilizations in our galaxy with which communication might be possible
$R_* =$ the average rate of star formation in our galaxy
$f_\mathrm{p} =$ the fraction of those stars that have planets
$n_\mathrm{e} =$ the average number of planets that can potentially support life per star that has planets
$f_\mathrm{l} =$ the fraction of planets that could support life that actually develop life at some point
$f_\mathrm{i} =$ the fraction of planets with life that actually go on to develop intelligent life
$f_\mathrm{c} =$ the fraction of civilizations that develop a technology that releases detectable signs of their existence into space
$L =$ the length of time for which such civilizations release detectable signals into space

LOOKING UP

As our own technology changes, so will the ways we hunt for our galactic neighbours. The Hubble Space Telescope has opened a window onto the universe. The Kepler space telescope has revealed thousands of new exoplanets, including planets in habitable *Goldilocks* zones around stars. The new James Webb space telescope, soon to be launched, will be able to detect not just more exoplanets, but also the chemical signatures of their atmospheres, which will reveal much about the activities of anything that lives there.

Keep an eye out for large-scale alien technology and engineering – alien mega-structures orbiting stars, such as heat signatures from Dyson spheres, that may be built by advanced civilizations to harness the star's power. The technological progress of a civilization can be ranked on the Kardashev Scale, named after the Russian astrophysicist Nikolai Kardashev, which describes a civilization's ability to harness energy from its parent star or galaxy. There are three levels: Types I to III. Any civilization able to abduct you is going to be at least a Type II, able to harness the total energy of the parent star. We're not even on the scale yet.

LOCATION

Wherever you are in the world, there will have been some UFO sightings nearby: Area 51 and Roswell in America of course. Mexico City's mass sighting, Rendlesham Forest in England and Bonnybridge in Scotland are a few of the most notorious. While there have been many en masse UFO sightings, abductions seem to occur at night in isolated places and when there are few people to witness the event. Try dark forests, lonesome roads with no traffic, isolated farms. Or hang around your local top-secret military base.

LANGUAGE

It's always useful to have a few local phrases up your sleeve. Music is a universal feature of all civilizations. Is it reasonable to extend this concept to civilizations beyond earth, as Carl Sagan and his Voyager golden record team did? Music as language has become a common motif in our science fiction, from the swanee-whistling Clangers, to Spielberg's *Close Encounters of the Third Kind*, where the music educationalist John Curwen's solfège 'do-re-mi' hand gestures were used to teach the visiting musical spacecraft's tonal messages.

Domingo Gonsales learned the musical language of the 'Lunas' during his sojourn on the 'moone': 'The Difficulty of that language is not to bee conceived, and the reasons thereof are especially two: First, because it hath no affinitie with any other that ever I heard. Secondly, because it consisteth not so much of words and Letters, as of tunes and uncouth sounds that no letters can expresse.'

Here are a couple of useful phrases from the Lunas' musical language that he picked up for you to try:

Translation: 'Glorie be to God Alone'

Translation: 'Gonsales'

BE PREPARED

If you get abducted then you won't know where you may end up. Will they have Wi-Fi or snacks on board? Make sure your phone is charged, with a good data-roaming plan and lots of hard-drive space for taking videos and pictures. Have a small backpack ready with warm clothes. There's a good chance you may be 'examined', so clean underwear is a must. A pen and paper, some water and food. Prepare for the unexpected. Check with your travel insurance company if leaving the planet is included. If not, it's worth getting in touch with a specialist.

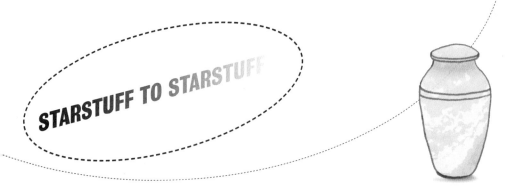

HOW TO LEAVE THE PLANET WHEN YOU'RE DEAD

There are those in life who are happy to wait for their ship to come in. There are those who have the drive to swim out and meet their ship. And those who have the vision and resources not to wait for either, but rather build the ship themselves. But if you miss the boat altogether, it's worth remembering that leaving the planet is a lot cheaper and more straightforward when you're dead. Your expansive and fragile 75 kg body will be a lot more compact and resilient, and will weigh only a couple of kilograms when it's reduced to ash. And as any rocket engineer will tell you, when it comes to payloads, weight and size are everything.

There are several companies who have spotted this niche and are now in the business of returning you back to the immeasurable heavens from whence you came – companies with tasteful names like *Ascending Memories*, *Elysium Space*, *Ascension* and *Celestis*, who offer to send your ashes into space with something for every budget. Sub-orbital? Orbital? Moon? Deep space? The wallet's the limit. Starting at a couple of thousand dollars, you (or a least a spoonful of 'you') can live forever among the stars. The view of the universe might not be as transformative, but like death,

the experience will last forever – or at least until your spacecraft's orbit decays and you return to earth re-cremated as a shooting star. Star Trek creator Gene Roddenberry, 1960s counterculture icon Timothy Leary and a host of other dead notables are all satisfied customers of this celestial burial at space. It took a couple of attempts for Star Trek actor James Doohan to make the trip – some of his ashes were launched on a SpaceX Falcon 1 rocket in 2008 that failed to make it into orbit, but he finally made it aboard a Falcon 9 in 2012, beaming up in the company of Project Mercury astronaut Gordon Cooper. The planetary scientist Eugene Shoemaker went one step further – his remains were sent crashing into the moon on board the Lunar Prospector to mingle with the moondust. If you want to go for your own *memento mori* moonshot, try to make sure your death coincides with a launch to avoid disappointment. Right now *Celestis* are offering to send a portion of your remains to the moon for $12,500 on board its 'Luna 02', the Lunar XPRIZE Moon Express spacecraft. Launch date: to be confirmed. It might be worth planning ahead.

Frustratingly, our *restless sphere* has made it hard for those with aspiring minds to physically stray too far. Hard, but not impossible. Whether you're dead or alive, when it comes to leaving the planet, history has taught us that timing is crucial: what's needed is a moment of synchronicity where the right technology, the right politics, the right price and the right planets all align. Whether you run a national space agency, or you're a jobbing astronaut, or a ten-year-old girl staring back in time through the bedroom window at the night sky, being in the right place at the right time is everything. Where does that leave you?

Our time down here is fleeting. There are no guarantees of anything. But if you're persistent, if you use your time creatively, and if you're dealt a lucky hand along the way, the stubborn window might creak open to let you through. Although the golden rule, as any space traveller caught without their spacesuit will tell you, is *don't hold your breath*. In the meantime, you can at least keep the window clean and enjoy the view. Allow yourself to feel the tug of the strings and the sound of the beating wings of the *Gansa* pulling you higher and faster. With every journey, the real fun is in the planning – the anticipation of the grand adventure to come. And every grand adventure is launched by asking the most important question of all – *Well, shall we go?* To which the answer must be, of course, *Yes, let's go*.

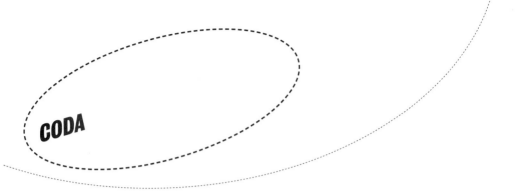

CODA

The Moth and the Star
by James Thurber

A young and impressionable moth once set his heart on a certain star. He told his mother about this and she counselled him to set his heart on a bridge lamp instead. 'Stars aren't the thing to hang around,' she said; 'lamps are the thing to hang around.' 'You get somewhere that way,' said the moth's father. 'You don't get anywhere chasing stars.' But the moth would not heed the words of either parent. Every evening at dusk when the star came out he would start flying toward it and every morning at dawn he would crawl back home worn out with his vain endeavour. One day his father said to him, 'You haven't burned a wing in months, boy, and it looks to me as if you were never going to. All your brothers have been badly burned flying around street lamps and all your sisters have been terribly singed flying around house lamps. Come on, now, get out of here and get yourself scorched! A big strapping moth like you without a mark on him!'

The moth left his father's house, but he would not fly around street lamps and he would not fly around house lamps. He went right on trying to reach the star, which was four and one-third light years, or twenty-five trillion miles, away. The moth thought it was just caught in the top branches of an elm. He never did reach the star, but he went right on trying, night after night, and when he was a very, very old moth he began to think that he really had reached the star and he went around saying so. This gave him a deep and lasting pleasure, and he lived to a great old age. His parents and his brothers and his sisters had all been burned to death when they were quite young.

Moral: Who flies afar from the sphere of our sorrow is here today and here tomorrow.

Opposite: **Sketch by Konstantin Tsiolkovsky (1933)**

27. Смотря наружу через стеклянн(ое) окно. ч?

Что видят.

1. Черное небо, усеянное разно-
цветными точками (звездами).
То же и обыкновенная, но
более яркая луна.
2. Сиреватое ослепительное солнце.
3. Земля занимает почти полнеба
ярко светит, и делает из ночи день.
4. Фазисы.

'Nature, that fram'd us of four elements
Warring within our breasts for regiment,
Doth teach us all to have aspiring minds:
Our souls, whose faculties can comprehend
The wondrous architecture of the world,
And measure every wandering planet's course,
Still climbing after knowledge infinite,
And always moving as the restless spheres,
Will us to wear ourselves, and never rest,
Until we reach the ripest fruit of all,
That perfect bliss and sole felicity,
The sweet fruition of an earthly crown.'

TAMBURLAINE THE GREAT
by Christoper Marlowe

A SHORT READING AND FILM LIST

BOOKS

THE MAN IN THE MOONE:
Francis Godwin (1638),
edited by William Poole (2009)

FROM THE EARTH TO THE MOON:
Jules Verne (1865)

NASA VOYAGER 1 & 2:
OWNER'S WORKSHOP MANUAL:
Christopher Riley, Richard Corfield and
Philip Dolling (2015)

SOYUZ: OWNER'S WORKSHOP MANUAL:
David Baker (2014)

ANIMALS IN SPACE: FROM RESEARCH ROCKETS
TO THE SPACE SHUTTLE:
Colin Burgess and Chris Dubbs (2007)

SOVIET SPACE DOGS:
Olesya Turkina (2014)

SELECTING THE MERCURY SEVEN:
Colin Burgess (2011)

EXPLORING SPACE:
A LADYBIRD ACHIEVEMENTS BOOK:
Roy Worvill, illustrated by B. Knight (1964)

CHOOSING THE RIGHT STUFF:
Patricia Santy (1994)

RIGHT STUFF, WRONG SEX:
AMERICA'S FIRST WOMEN IN SPACE PROGRAM:
Margaret A. Weitekamp (2004)

THE RIGHT STUFF:
Tom Wolfe (1979)

SPACESUITS: THE SMITHSONIAN NATIONAL AIR
AND SPACE MUSEUM COLLECTION:
Amanda Young (2009)

SPACESUIT: FASHIONING APOLLO:
Nicholas De Monchaux (2011)

U.S. SPACE GEAR:
Lillian D. Kozloski (1994)

ROCKET DREAMS:
Marina Benjamin (2003)

DRAGONFLY:
NASA AND THE CRISIS ABOARD MIR:
Bryan Burrough (1998)

LOST MOON:
Jim Lovell and Jeffrey Kluger (1994)

MURMURS OF EARTH: THE VOYAGER
INTERSTELLAR RECORD:
Carl Sagan (1978)

FILMS

BANG GOES THE THEORY:
BBC1 popular science series (2009-2014)

COSMONAUTS:
HOW RUSSIA WON THE SPACE RACE:
BBC documentary, directed by Michael Lachmann
(first broadcast 13 October 2014)

THE ENGINES THAT CAME IN FROM THE COLD:
Channel 4 documentary from the Equinox Series,
directed by Dan Clifton (first broadcast 1 March 2001)

FRAU IM MOND:
Film directed by Fritz Lang (1929)

MAN IN SPACE:
Disney documentary, directed by Ward Kimball
(first broadcast 9 March 1955)

THE RIGHT STUFF:
Film based on Tom Wolfe's famous book,
directed by Philip Kaufman (1983)

ORPHANS OF APOLLO:
Film directed by Becky Neiman and
Michael Potter (2008)

MIR MORTALS:
BBC Horizon, produced by Jill Fullerton-Smith
(first broadcast 23 April 1998)

VOYAGER: TO THE FINAL FRONTIER:
BBC documentary directed by Christopher Riley
(first broadcast 24 October 2012)

THE RESTLESS SPHERE: THE STORY OF
INTERNATIONAL GEOPHYSICAL YEAR:
BBC documentary, presented by
HRH Prince Phillip The Duke of Edinburgh
(first broadcast 30 June 1957)

NOTES AND SUNDRIES

Each paragraph in this book is worthy of writing a whole book in itself. And many have done so. Here are a few more places I went to, which will expand the areas on which I've only briefly touched.

These notes were compiled with the help of Tamsin Edwards. All webpages were last accessed July 2017.

PREFACE

1. 'EARTH IN THE VIEWPORT …': *The Grass Beside Our House* by Zemlyane (1979). With grateful thanks to Svetlana Ivanova Poperechnaya for her kind permission to reproduce the lyrics of Anatoly Poperechny, and to Dr Iya Whiteley for agreeing to call her from the green room of the Cheltenham Science Festival. Translation by George Ross (2010), from russongs.tumblr.com/post/73998256115/zemlyane-the-grass-beside-our-home.

2. 553 HUMANS HAVE DONE IT: As of July 2017, according to the official FAI (Fédération Aéronautique Internationale) definition, or 559 according to the United States Air Force definition. See https://www.worldspaceflight.com/bios/stats.php.

3. EIGHT PEOPLE HAVE DIED EN-ROUTE TO SPACE AND 11 ON THE WAY HOME: Those people were Michael Smith, Dick Scobee, Ronald McNair, Ellison Onizuka, Christa McAuliffe, Gregory Jarvis, and Judith Resnik (Challenger, 1986); Michael Alsbury (SpaceShipTwo, 2014); Vladimir Komarov (Soyuz 1, 1967); Georgy Dobrovolsky, Vladislav Volkov, and Viktor Patsayev (Soyuz 11, 1971); Rick D. Husband, William McCool, Michael Anderson, Kalpana Chawla, David Brown, Laurel Clark and Ilan Ramon (Columbia, 2003). See 'Health and safety' chapter for some of those lost in training.

LET'S GO

4. 'LET'S GO ON A JOURNEY INTO SPACE…': Based on a conversation with astrophysicist Dr Sheona Urquhart.

5. JUNK BORN FROM JUNK AD INFINITUM: Inspired by the beautiful *Project Adrift*, via twitter.com/FengyunAdrift/status/826408791038103553. See 'Adventures in Low Earth Orbit' chapter for more.

FIRST STAGE: GROUND CONTROL

WHAT'S HOLDING YOU BACK?

THE JACOBEAN SPACE PROGRAMME

6. THE JACOBEAN SPACE PROGRAMME: A phrase coined by Prof Allan Chapman, for example in his 11 October 2004 talk *The Jacobean Space Programme – Wings, Springs and Gunpowder:*

Flying to the Moon from 17th Century England at Gresham College. Available at www.gresham.ac.uk/lectures-and-events/the-jacobean-space-programme-wings-springs-and-gunpowder-flying-to-the-moon-from.

7. THAT PULLED HIM ACROSS THE CIRCUMLUNAR SPACE …IN SOME 12 DAYS: For the mathematics of goose flight, see Andrew J. Simoson (2007) *Pursuit Curves for the Man in the Moone*, The College Mathematics Journal, vol. 38, 330-338. Available at www.maa.org/sites/default/files/pdf/upload_library/22/Polya/simoson330.pdf.

8. 'CELESTIAL DON QUIXOTE': Prof Allan Chapman again, as above.

THE RENAISSANCE RIGHT STUFF

9. 'WE DO NOT ADMIT DESK-BOUND HUMANS…': From *VIII: The Daemon from Levania* in *Somnium* (1608). Translated by Tom Metcalfe for The Somnium Project. Available at somniumproject.wordpress.com/somnium/viii.

TRUE STORIES

10. 'MY SUBJECT IS, THEN, WHAT I HAVE NEITHER SEEN…': The *Lucian of Samosata Project* has many public domain translations. This one is by H.W. Fowler and F. G. Oxford, The Clarendon Press. 1905. Available at lucianofsamosata.info.

11. '1) BY SPIRITS, OR ANGELS…': From the start of Chapter VII, *Concerning the Art of Flying. The several ways thereby this hath been, or may be attempted*, in Book II, *Daedalus of Mathematical Magick* (1648). Also known as: *Mathematical Magic, or, The Wonders that May By Performed by Mechanical Geometry. In Two Books. Concerning Mechanical { Powers. Motions.} Being one of the most easy, pleasant, useful (and yet most neglected) Part of the Mathematics. Not before treated of in this Language.* Which is rather nice.

CROSSING THE LINE

12. VON KÁRMÁN ESTABLISHED THE DEMARCATION LINE: See 100 km *Altitude Boundary for Astronautics* by Dr Sanz Fernández de Córdoba, on the FAI website. Available at www.fai.org/icare-records/100km-altitude-boundary-for-astronauts.

GRAVITY WELLS

13. FOR THOSE ATTEMPTING TO DEFY 'GRAVITAS': From p123 of Allan Chapman (1991) *'A World in the Moon': John Wilkins and his Lunar Voyage of 1640*, Quarterly Journal of the Royal Astronomical Society, vol. 32, 121–132. Available at adsabs.harvard.edu/full/1991QJRAS..32..121C.

14. POTATO RADIUS: An estimate published in conference proceedings by Charles Lineweaver and Marc Norman (2010) *The Potato Radius: a Lower Minimum Size for Dwarf Planets*, Australian Space Science Conference Series: Proceedings

of the 9th Australian Space Science Conference, edited by Short and Cairns, National Space Society of Australia, April. Available at arxiv.org/abs/1004.1091.

15. HAMMER AND FEATHER: For details of the mission profile, see Al's website. Available at alworden.com.

16. HAFELE-KEATING: In this experiment, the effect is actually about half due to General Relativity (gravity of the earth) and half to Special Relativity (going really fast).

FALLING AROUND THE EARTH

17. 'HOW ABOUT THAT? MR GALILEO WAS CORRECT IN HIS FINDINGS': Dave Scott (1971), NASA Apollo 15 film archive. Available at nssdc. gsfc.nasa.gov/planetary/lunar/apollo_15_feather_drop.html.

HOW TO BUILD A ROCKET

18. 'WHEN YOU WANT TO BUILD A SHIP...': Widely attributed to Antoine de Saint Exupéry, but it's a mistranslation in a particular edition of *Citadelle* (1948). This rather nice variation is from the Antoine de Saint Exupéry section of Wikiquote.

19. 'DESTINATION: THE MOON, OR MOSCOW. THE PLANETS, OR PEKING...': James Burke in *Eat Drink and Be Merry*, episode 8 of the BBC documentary series *Connections*. First transmission 5 December 1978.

1. THE RUSSIAN MYSTIC VISIONARY OF ROCKETRY

20. 'THE EARTH IS THE CRADLE OF HUMANITY...': From *The Exploration of the Universe with Reaction Engines* (1911), published in *Herald of Aeronautics*. This translation was used in the wonderful Science Museum exhibition *Cosmonauts: Birth of the Space Age* in 2015/6.

21. 'FIRST, INEVITABLY, THE IDEA, THE FANTASY...': Source of translation unknown.

2. THE AMERICAN BACKYARD ROCKET ENGINEER WHO LIKED TO CLIMB TREES

22. MORE A MAN OF ACTION: Robert H. Goddard made the case for transporting humans by rocket in ROCKETING to the Moon, p47 of Modern Mechanix, January 1930.

23. 'SET OFF RED FIRE AT A PREARRANGED TIME...': From his diary, cited in David P. Stern (1999) *Remembering Robert Goddard's Vision 100 Years Later*, Eos, vol. 80, issue 38, 441–2. doi: 10.1029/99EO00322. Available at onlinelibrary.wiley.com/doi/10.1029/99EO00322/full.

24. 'THE GREAT ROCKET...': From ROCKETING to the Moon, as above.

3. THE CHIEF DESIGNER

25. OFFERED THE NOBEL PRIZE TWICE: According to Korolev's daughter interviewed in *The Red Stuff* (2000), directed by Leo De Boer, Netherlands, Pieter Van Huystee Film and Television. 'Everything was so secret that my father was even denied the Nobel Prize twice. The Nobel Committee had approached our government to give him the Nobel Prize for the Sputnik and later for Gagarin. Khrushchev then said: "Our whole people has created these techniques." He did not win the Nobel Prize.' Available at www.youtube.com/watch?v=QQL9kUCdsu4 [31:00].

4. HERMANN OBERTH, A WOMAN IN THE 'MOONE' AND THE NAZI FLYING SAUCERS

26. '"NEVER" DOES NOT EXIST FOR THE HUMAN MIND...': The opening title card of Fritz Lang's *Frau im Mond* (1929), Universum Film. Reproduced with permission of Murnau-Stiftung.

27. NAZI FLYING SAUCERS: See Hermann Oberth (1955) *They Come From Outer Space*, Flying Saucer Review, vol. 1, no. 2, May-June.

5. DR SPACE AND THE VENGEANCE WEAPON

28. 'THIS THIRD DAY OF OCTOBER, 1942, IS THE FIRST OF A NEW ERA...': In a speech at Peenemünde.

29. WERNHER VON BRAUN: Sheet music via MusicNotes.com. Reproduced with the very kind and charming permission of Tom Lehrer.

YOUR FUTURE RIDE TO SPACE

30. UPWARDS OF $5 MILLION: The classic number quoted for the Space Shuttle is $10,000 per pound ($22,000 per kg), which would be $10M. Estimates for more current and future rockets abound, but Harry W. Jones (2015) gives a potential range for the SLS of $7140 per kg to double that, so $3.2M to $6.4M for the paint job, in *Estimating the Life Cycle Cost of Space Systems*, ICES-2015-041, 45th International Conference on Environmental Systems, 12-16 July 2015, Bellevue, Washington. Available at ntrs.nasa.gov/search.jsp?R=20160001190.

CAN I TAKE MY DOG?

31. 'ALL THE UNIVERSE IS FULL OF THE LIVES OF PERFECT CREATURES...': From *The Scientific Ethics* (1930). Source of translation unknown.

32. 'ARDAN WISHED TO CONVEY A NUMBER OF ANIMALS OF DIFFERENT SORTS...': From p291-292 of my copy of Jules Verne (1865), *From the Earth to the Moon*, in: *Around the World in Eighty Days; From The Earth to the Moon Direct; 20,000 Leagues Under the Sea*, Octopus Books Limited, London (this edition 1978).

33. 'WEIGHTLESS RABBIT FLIES...': From *The Marfusha Haiku*, written and read by John Talley-Jones. Available at www.youtube.com/watch?v=sU_fqMeaLQw.

34. LIVING OUT ITS DAYS IN THE PERSONAL ZOO OF MARIE ANTOINETTE: From p12 of S.L. Kotar and J.E. Gessler (2011), *Ballooning: A History, 1782-1900*, McFarland, Jefferson, North Carolina, USA.

MOSS PIGLETS

35. A 400G RAT WILL COST YOU AROUND $9000 TO GET INTO SPACE: Estimate of $8800 for an average weight rat by commenter 'Arcane' using classic NASA payload estimate above, under Kelsey Campbell-Dollaghan (2014) *NASA's Space Rats Are Checking Into This Swanky New Rodent*

Hotel, Gizmodo, 27 May. Available at gizmodo. com/nasas-space-rats-are-checking-into-this-swanky-new-rode-1582002930. But some rockets might cost you less, some more.

36. MANY SURVIVED AND WENT ON TO REPRODUCE: K. Ingemar Jönsson and others (2008) *Tardigrades survive exposure to space in low Earth orbit.* Current Biology, vol. 18, 729–731. Available at www.sciencedirect.com/science/article/pii/ S0960982208008051.

BLOSSOMING – FRUIT FLIES

37. PROJECT BLOSSOM: See *Project Blossom 1 - AMC Pictorial Review 1 V-2 Rocket* from the Periscope Film LLC archive. Uploaded online 30 August 2014. Available at www.youtube.com/ watch?v=A5PWjS131cs.

38. THE FLIES RETURNED SAFE AND SOUND: From p31 of Colin Burgess and Chris Dubbs (2007) *Animals in Space: From Research Rockets to the Space Shuttle*, Springer.

39. YOUR FLIES ARE 728 GENERATIONS AWAY: According to Dr Adam Rutherford's back of the envelope reply to my flippant question: 'Average annual temperature in Santa Fe is only 10°C, which increases *Drosophila melanogaster* development time to a whopping 50 days, so life cycle is much longer. But let's say 5 weeks, as they won't have stayed in Santa Fe for 70 years. So that makes around 728 generations.'

SATELLITE DOGS

40. SATELLITE DOGS: This phrase taken from p203 of Burgess and Dubbs, *Animals in Space*, as above.

41. 'THE DOG WILL DIE, WE CAN'T SAVE IT': The Daily Mirror (1957), as reported by TIME Magazine (1957) ANIMALS: *The She-Hound of Heaven*, 18 November. Available from content.time.com/ time/subscriber/article/0,33009,868018,00.html.

42. 'LAIKA, SWEET LOYAL LAIKA…': From *Tyapa, Borka and the Rocket*, a 1962 story book for children by Marta Baranova and Yevgeny Veltisov, quoted on p111 of Olesya Turkina's beautiful book *Soviet Space Dogs* (2014), Fuel, London.

43. THE SPACE TO COCK THEIR LEGS: From p107 of *Soviet Space Dogs*, as above.

44. SUFFERING FROM MASSIVE STRESS AND OVERHEATING …BY THE THIRD OR FOURTH ORBIT: D.C. Malashenkov (2002), *Some Unknown Pages of the Living Organisms' First Orbital Flight*, IAF (International Astronautical Federation) abstracts, 34th COSPAR Scientific Assembly, The Second World Space Congress, held 10-19 October, 2002 in Houston, TX, USA., p.IAA-2-2-05. Available at adsabs.harvard.edu/abs/2002iaf.. confE.288M

45. ANOTHER 2570 REVOLUTIONS OF THE EARTH …LAIKA'S RETURN HOME: From p164 of Burgess and Dubbs, *Animals in Space*, as above.

46. STRELKA'S PUPPIES: From p207 of Burgess and Dubbs, *Animals in Space*, as above.

47. 48 DOGS HAD BEEN LAUNCHED. 20 HAD DIED. ALL WERE LOVED: BBC4, *Space Dogs*, www. bbc.co.uk/programmes/b00lkvtp. Available via UKAstronomy at www.youtube.com/ watch?v=cL03G46J-CM.

HAM – PRELUDE TO MAN

48. PROJECT ALBERT: From p40 onwards of Burgess and Dubbs, *Animals in Space*, as above.

49. FRUIT FLY LARVAE, HUMAN BLOOD …AND SPERM: From p136 of Burgess and Dubbs, *Animals in Space*, as above.

50. HAM HAD HIS PRE-FLIGHT BREAKFAST: From p249 of Burgess and Dubbs, *Animals in Space*, as above.

51. HAM'S POST-FLIGHT CELEBRITY AND CULTURAL ICONIC STATUS: There are mentions of TV appearances, but he definitely didn't star in a movie with Evil Knievel, despite the references to this in many books due to a misunderstanding of an old internet article.

52. ENOS' UNFORTUNATE NICKNAME: According to Mary Roach, who cleared his name in her 2010 book *Packing for Mars: The Curious Science of Life in the Void*, W. W. Norton & Company.

53. 'IN ADDITION, THE SUBJECT HAD BROKEN THROUGH…': From p54 of James P. Henry and John D. Mosely [editors] (1963) *Results of the Project Mercury Ballistic and Orbital Chimpanzee Flights: National Aeronautics and Space Administration*, NASA Special Publication SP-39, Washington. Available at history.nasa.gov/ SP39Chimpanzee.pdf.

54. HAM'S REMAINS: Henry Nicholls (2013) *Ham the astrochimp: hero or victim?* The Guardian, 16 December. Available at www.theguardian. com/science/animal-magic/2013/dec/16/ham-chimpanzee-hero-or-victim.

ASTROCHAT AND ASTRORAT

55. 'HARDLY HAD THE SHELL BEEN OPENED…': From p282 of Jules Verne, *From the Earth to the Moon*, as above.

56. HAVING BEEN WELL TRAINED ON THE RAT CENTRIFUGE: You can see a video of Hector training in the short video *Hector, rat français de l'espace*, Institut national de l'audiovisuel (INA), 18 January 1961. Available at www.ina.fr/video/ AFE85008962.

IVAN IVANOVICH – THE RUSSIAN DOLL

57. 'PHANTOM COSMONAUT': Megan Garber (2013) *The Doll That Helped the Soviets Beat the U.S. to Space*, The Atlantic, 28 March. Available at www. theatlantic.com/technology/archive/2013/03/ the-doll-that-helped-the-soviets-beat-the-us-to-space/274400.

58. EIGHTY MICE, GUINEA PIGS [AND] A RECIPE FOR BORSCHT: Megan Garber, *The Doll That Helped the Soviets Beat the U.S. to Space*, as above.

ZOND 5 – THE TORTOISES WHO THOUGHT THE WORLD WAS FLAT AND THEN SAW IT WAS ROUND

59. INTERCEPTED A MYSTERIOUS HUMAN VOICE: Sven Grahn (2008) *Jodrell Bank's role in early space tracking activities - Part 2*. Available at www. jb.man.ac.uk/history/tracking/part2.html

60. RUSSIAN COSMONAUT VALERY BYKOVSKY: From p186 of Brian Harvey (2007) *Soviet and Russian Lunar Exploration*, Springer Praxis, Chichester, UK.

61. ANOTHER 70KG HUMAN MANNEQUIN… A PAIR

OF RUSSIAN STEPPE TORTOISES: NASA (2017) *Zond 5*, 21 March. Available at nssdc.gsfc.nasa.gov/nmc/spacecraftDisplay.do?id=1968-076A.

HOW TO GO INTO SPACE WITHOUT LEAVING HOME

62. 'A PERSON OR THING SEEN AS COMPARABLE TO ANOTHER...': Oxford Dictionary.

MOON GOOSE ANALOGUE: LUNAR MIGRATION BIRD FACILITY

63. 'WHAT HAPPENED TO THE MOON GEESE IN THE TWENTY FIRST CENTURY?': Agnes Meyer-Brandis in THE MOON GOOSE ANALOGUE - *documentation*, 22 March 2012. Available at vimeo.com/38986659. For more on this project, commissioned by Arts Catalyst and FACT Liverpool, see www.blubblubb.net/mga. Visiting this extraordinary installation in 2014 was, in many ways, the seed of this book. Thank you Agnes.

SPACE STAYCATION

64. SAGAN'S ORANGE JACKET: Jon Lomberg (see 'Envoys' chapter) told me this story over dinner with Chris Riley.

LONG DURATION MISSIONS

65. HAND-WRITTEN LETTER TO THE MINERS: Romain Charles of the Mars 500 project told me this story at the European Astronaut Centre, 3 March 2017.

WHITE MARS

66. 'POLAR EXPLORATION IS AT ONCE...': From page xlix of Apsley Cherry-Garrard (1922), *The Worst Journey in the World*, Picador, London (this edition 2001). Another important book in my life.
67. THE BEST PLACE ON EARTH FOR ASTRONOMICAL OBSERVATIONS: According to Will Saunders and others (2009) *Where Is the Best Site on Earth? Domes A, B, C, and F, and Ridges A and B*, Publications of the Astronomical Society of the Pacific, vol. 121, no. 883, 976–992. Available at iopscience.iop.org/article/10.1086/605780.
68. WHERE THE 'STRATOSPHERE COMES TO THE GROUND': From Steward Observatory Radio Astronomy Laboratory (2013) *The High Elevation Antarctic Terahertz Telescope*. Available at soral.as.arizona.edu/HEAT.

TRAVEL GUIDE: THE HIVERNAUT

69. BETH HEALEY: Interviewed at the Royal Geographical Society, 31 January 2017. Thanks for the many space chats Beth.

SECOND STAGE: CLEARING THE TOWER

70. 'BETWEEN YOU AND INFINITY THERE ARE BILLIONS OF STARS': From p87 of Leonard de Vries (1963) *The Second Book Of Experiments*, Jarrold & Sons Ltd, Norwich. Translated by Eric G. Breeze. A writer (and translator) who continues to take my breath away. A children's book that takes beauty and wonder seriously.

DO I HAVE THE RIGHT STUFF?

AL WORDEN, APOLLO 15 COMMAND MODULE PILOT, ON THE MOMENT OF LIFTOFF ...

71. AL WORDEN: Interviewed at Alton Towers, 3 October 2016. Transcribed by Tamsin Edwards.
72. 'BETWEEN THE STIMULUS AND THE RESPONSE...': Frequently attributed to Viktor Frankl, but actually anonymous via Stephen R. Covey. See Franz Vesely, *Alleged quote: 'Between stimulus and response...'*, The Official Website of the Victor Frankl Institute Vienna. Available at www.viktorfrankl.org/e/quote_stimulus.html.
73. 'THAT INDEFINABLE, UNUTTERABLE INTEGRAL STUFF': From p30 of Tom Wolfe (1979), *The Right Stuff*, Vintage Books, London (this edition 2005).
74. 'DEAR GOD. PLEASE DON'T LET ME FUCK UP': Alan Shepard, as misquoted on p249 of Tom Wolfe's *The Right Stuff*, as above. He apparently said 'Don't fuck up Shepard', according to various sources, e.g. p34 of Mike Fuller (2014), *Our Beautiful Moon and its Mysterious Magnetism: A Long Voyage of Discovery*, Springer, London.

THE RIGHT ORIGAMI

75. 1000 ORIGAMI PAPER CRANES: From Mary Roach (2010) *The trials of the modern-day astronaut*, The Telegraph, 14 September. Available at www.telegraph.co.uk/news/science/7968400/The-trials-of-the-modern-day-astronaut.html. See also p277 of Carol Norberg [editor] (2013) *Human Spaceflight and Exploration*, Springer Praxis, Chichester. I practiced making these with Dr Iya Whiteley in the Cheltenham Science Festival green room.

THE RIGHT START

76. NASA OPENED ITS DOORS FOR BUSINESS: From p26 of Colin Burgess (2011), *Selecting the Mercury Seven*, Springer-Verlag, New York.
77. '1. TO SURVIVE...': From p45 of Mae Mills (1965), *Space Medicine In Project Mercury*, NASA Special Publication-4003. Available at history.nasa.gov/SP-4003.pdf.

THE RIGHT SEARCH

78. 1500 HOURS FLIGHT TIME ...AND HOLD A UNIVERSITY DEGREE: From p45–47 of Mae Mills, *Space Medicine In Project Mercury*, as above.
79. 32 PASSED ...AND WERE DISPATCHED TO THE INFAMOUS LOVELACE CLINIC: From p48 of Mae Mills, *Space Medicine In Project Mercury*, as above.

THE RIGHT CLINIC

80. HIS LEFT HAND ...FROZE INSTANTLY: From p69 of LIFE (1943) *Army Doctor's Record Parachute Jump*, 9 August.
81. HE BROKE THE WORLD PARACHUTE ALTITUDE RECORD: From p29 of Margaret A. Weitekamp (2004) *Right Stuff, Wrong Sex: America's First Women in Space Program*, The John Hopkins University press, Baltimore, Maryland, USA.

THE RIGHT TESTS

82. 'RIDING THE STEEL EEL': From p214 of Colin Burgess, *Selecting the Mercury Seven*, as above.
83. EVEN A SPERM SAMPLE HAD TO BE GIVEN:

From p216 of Colin Burgess, *Selecting the Mercury Seven*, as above.

84. 'YOU WISHED TO KNOW ALL ABOUT MY GRANDFATHER...': See for example p217 of Colin Burgess, *Selecting the Mercury Seven*, as above.

85. THE 31 CANDIDATES ...WERE NOW ASSIGNED LETTERS: From p233 of Colin Burgess, *Selecting the Mercury Seven*, as above.

86. STRESS TESTS: Mostly taken from Charles L. Wilson (1959) *Project Mercury Candidate Evaluation Program*, Wright Air Development Center Technical Report 59-505. Available at www.dtic.mil/dtic/tr/fulltext/u2/234749.pdf.

87. A HEAT CHAMBER TEST ...WITH A RECTAL THERMOMETER: From p238 of Colin Burgess, *Selecting the Mercury Seven*, as above.

88. THE 'IDIOT BOX': From p243 of Colin Burgess, *Selecting the Mercury Seven*, as above.

89. PSYCHOLOGICAL TESTS: From p83 of Charles L. Wilson (1959), *Project Mercury Candidate Evaluation Program*, as above.

90. FROM 508: From p260–1 of Colin Burgess, *Selecting the Mercury Seven*, as above.

91. 'OUTSTANDING, WITHOUT RESERVATIONS': From p89 of Charles L. Wilson (1959) *Project Mercury Candidate Evaluation Program*, as above. In fact only 6 of these 7 were chosen. The 7th was from the list of three candidates judged 'Outstanding with reservations'. Perhaps it was the candidate for whom the reservation was not medical but that he 'was not entirely sure that he desired to continue on in Project Mercury'.

92. '...PSYCHOLOGICALLY HEALTHY MEN...': From p265 of Patricia Santy (1994), *Choosing the Right Stuff*. Source for quote given as George Ruff on p241.

'SHOULD A GIRL BE FIRST IN SPACE?'

93. '...A FLAT CHESTED LIGHTWEIGHT...' From p119 of my copy of Look (1960), *Should A Girl be First in Space?* 2 February, vol. 24, no. 3, p112–119.

94. 'EXPERTS IN OUR SPACE PROGRAM PREDICT WOMEN WILL BE CONSIDERED...': From p116 of Look (1960), *Should A Girl be First in Space?*, as above.

95. 'ASTRONAUTRIX', 'FEMINAUT', 'ASTRONETTE': From p78 of Margaret A. Weitekamp, *Right Stuff, Wrong Sex America's First Women in Space Program*, as above.

96. WOMEN HAD VARIOUS PHYSIOLOGICAL ADVANTAGES: From p64–65 of Margaret A. Weitekamp, *Right Stuff, Wrong Sex America's First Women in Space Program*, as above.

97. JERRIE COBB STORY: From p76–77 of Margaret A. Weitekamp, *Right Stuff, Wrong Sex America's First Women in Space Program*, as above.

98. 'BEYOND EXPECTATIONS': From p77 of Margaret A. Weitekamp, *Right Stuff, Wrong Sex America's First Women in Space Program*, as above.

99. TWENTY-FIVE WOMEN COMPLETED THE LOVELACE TESTS, WITH THIRTEEN PASSING: From p95 of Margaret A. Weitekamp, *Right Stuff, Wrong Sex America's First Women in Space Program*, as above.

100. THE FIRST LADY ASTRONAUT TRAINEES (FLATS) OR THE MERCURY 13: From p178-9 of Margaret A. Weitekamp, *Right Stuff, Wrong Sex America's First Women in Space Program*, as above.

AM I AN ASTRONAUT?

101. NASA CANDIDATE QUALIFICATION REQUIREMENTS: Based on NASA (2011) *Astronaut Selection and Training*, FS-2011-11-057-JSC. Available at www.nasa.gov/centers/johnson/pdf/606877main_FS-2011-11-057-JSC-astro_trng.pdf. And also *Astronaut Candidate Program*, applications open 14 December, 2015. Available at astronauts.nasa.gov/content/broch00.htm.

TRAVEL GUIDE:
THE STEELY EYED MISSILE MAN

102. '...THE FLIGHT RULES DICTATED AN ABORT...': From p158 of Jim Lovell and Jeffrey Kluger (1994), *Lost Moon*, Houghton Mifflin Company, New York, USA.

THE AFRONAUTS

103. 'ZAMBIANS ARE INFERIOR TO NO MEN...': Edward Makuka Nkoloso (1964) *We're going to Mars! WITH A SPACEGIRL, TWO CATS AND A MISSIONARY*, as reported by Lusaka Times (2011) *Zambia's forgotten Space Program*, 28 January. Available at www.lusakatimes.com/2011/01/28/space-program.

104. AFRONAUT STORY: See Namwali Serpeli (2017) *The Zambian 'Afronaut' Who Wanted to Join the Space Race*, 11 March. Available at www.newyorker.com/culture/culture-desk/the-zambian-afronaut-who-wanted-to-join-the-space-race.

105. 'WHERE FATE AND HUMAN GLORY LEAD...': From Peter Collett (2009) *Zambia boldly goes*, New Statesman, 30 July. Available at www.newstatesman.com/africa/2009/08/nkoloso-zambia-moon-astronauts.

106. 'BY PRIMITIVE NATIVES': Reported by Lusaka Times, *Zambia's forgotten Space Program*, as above.

107. 'LOOK AT THAT TREE....': From Peter Collett, *Zambia boldly goes*, as above,

WHAT SHALL I WEAR?

108. 'IT IS AN ORDINARY MAN...': Narration from NASA/ILC Dover promotional film, probably mid-1960s. Archive footage used in *Survival In The Skies* by Arrow Media on which I worked, commissioned by the Smithsonian Channel, broadcast on the Discovery Channel (UK) May-June 2017. Thanks to Chris Riley.

SUITS OF ARMOUR

109. A = APOLLO: From p99 of Lillian D. Kozloski (1994) *U.S. Space Gear: Outfitting the Astronaut*, Smithsonian Institution Press, Washington, USA.

110. STITCHING...BOXING GLOVES: From NASA (ca. 1970) *Clip from Moonwalk One, ca. 1970: Space Suit*, Headquarters' Films Relating to Aeronautics, US National Archives, uploaded 16 July 2014. Available at www.youtube.com/watch?v=4BlVRLTuCfU.

111. MICHELLE TICE, JULIA BROWN, DELEMA AUSTIN, DELORES ZEROLES, DORIS BOISEY, DELEMA COMEGYS: From p82, p83 and p89 of Lillian D. Kozloski, *U.S. Space Gear: Outfitting the Astronaut*, as above.

112. [WE LIVE] AT THE BOTTOM OF AN OCEAN OF

AIR: I have often used this phrase, but apparently Evangelista Toricelli used it first: 'Viviamo nel fundo d'un pelago d'aria elementare' in 1644, via Prof Andrea Sella in BBC Radio 4 *Mercury - Chemistry's Jekyll and Hyde*, first broadcast 1 May 2017. Available at www.bbc.co.uk/programmes/b08n2ltx.

UNDER PRESSURE

113. 'IT WAS NATURE'S CRUELLEST TORTURE....': From William R. Rankin (1960), *The Man Who Rode the Thunder*, pages 150–63, reprinted by permission of Simon & Schuster in Joseph J. Corn [editor] (2011) *Into the Blue: American Writers on Aviation and Spaceflight*, The Library of America.

114. THE FATE OF THE RUSSIAN CREW OF A SOYUZ SPACECRAFT: NASA (2010) *Descent into the Void*, System Failure Case Studies, vol. 4, issue 9, 1-4, September. Available at spaceflight.nasa.gov/outreach/SignificantIncidents/assets/descent-into-the-void.pdf

115. ATMOSPHERIC DATA TABLE: *U.S. Naval Flight Surgeon's Manual*, Naval Aerospace Medical Institute, Third Edition, 1991. From p19 of Dennis R. Jenkins (2012) *Dressing for altitude: U.S. aviation pressure suits – Wiley Post to space shuttle*, NASA. Available at www.nasa.gov/pdf/683215main_DressingAltitude-ebook.pdf.

116. YOU'VE GOT ABOUT TEN SECONDS OF CONSCIOUSNESS: From p5 of NASA (1973) [editors Parker Jr. and West] *Bioastronautics data book: second Edition*, NASA SP-3006. Available at ntrs.nasa.gov/search.jsp?R=19730006364.

SPACESUIT HISTORY

117. 'MY IDEA IS TO EMPLOY A SUIT, SOMETHING LIKE A DIVER'S OUTFIT...': From p493 of my copy of Wiley Post (1934) *Wiley Post Seeks New Record*, Popular Mechanics, October, 492-495.

118. 'THE FIRST TAILOR OF THE SPACE AGE': A phrase often used, for example on p64 of Colin Burgess (2016), *Aurora 7: The Mercury Space Flight of M. Scott Carpenter*, Springer International Publishing.

SILVER SUITS

119. SILVER BOOTS STORY: Various, e.g. Smithsonian (2017) *The Mercury 7 Caper*, Air & Space/Smithsonian Magazine, June. Available at www.airspacemag.com/as-next/mercury-7-caper-180963427.

EVA

120. LEONOV STORY: From p42 of David Baker (2014) *Soyuz: Owner's Workshop Manual*, Haynes Publishing, UK.

121. 'CAUCASUS! CAUCASUS!...': From *The Red Stuff*, as above.

122. FOR IDENTIFICATION ON LANDING: From p48 of Isaak P. Abramov and A. Ingemar Skoog (2003) *Russian Spacesuits*, Springer Praxis, Chichester.

A BETTER MAN THAN ME

123. TANG ...HAD LEAKED OUT: From p96 of Lillian D. Kozloski, *U.S. Space Gear: Outfitting the Astronaut*, as above. There are lots of stories about Tang in space.

124. 'THO' I'VE BELTED YOU AND FLAYED YOU...': Rudyard Kipling (1892) *Gunga Din*.

SPACE SUITS, THE NEXT GENERATION

125. 'SHE WILL NOT BE BOSOMY...': From p119 of *Look*, *Should A Girl be First in Space?*, as above.

126. MACES WITH A LIMITED DEGREE OF EVA CAPABILITY: Richard D. Watson (2014) *Modified Advanced Crew Escape Suit Intravehicular Activity Suit for Extravehicular Activity Mobility Evaluations*, 2014-ICES-194, 44th International Conference on Environmental Systems, 13-17 July 2014, Tucson, Arizona. Available at ntrs.nasa.gov/archive/nasa/casi.ntrs.nasa.gov/20140010572.pdf.

DO I NEED A VISA?

127. DR JILL STUART: by email and lots of fun conversations, February 2017. Thanks Jill.

SHOULD I BRING A PACKED LUNCH?

128. 'THEY DID SO, READILY ENOUGH, AND BROUGHT ME VERY GOOD FLESH AN FISH...': From Part IV of Francis Godwin (1638) *The Man in the Moone*. Quotes are taken from the public domain version, but I highly recommend the annotated edition by William Poole (2009), Broadview Press (this quote from p89).

HOW TO MAKE A SANDWICH IN SPACE

129. GEMINI III TRANSCRIPT: NASA (1965) *Gemini III Composite Air-to-Ground and Onboard Voice Tape Transcription*, April, 107 pages. Available at www.jsc.nasa.gov/history/mission_trans/gemini3.htm

LIVING OFF THE LAND

130. SPACE LETTUCE: Alan Yuhas (2015) *Nasa astronauts take first bites of lettuce grown in space: 'Tastes like arugula'*, The Guardian, 10 August. Available at www.theguardian.com/science/2015/aug/10/nasa-astronauts-lettuce-vegetables-grown-space.

131. POTATOCAM: Available at potatoes.space/mars.

132. SPACE ROCKS RECIPE: With huge thanks to Thorsten Schmidt. Website mallingschmidt.dk. Sorry for the endless emails. You are a true star.

HEALTH AND SAFETY

133. 'WHEN MAN STEPS INTO HIS ROCKET SHIP AND LEAVES THE EARTH BEHIND...': Walt Disney, *Man In Space*, directed by Ward Kimball. First broadcast on ABC on 9 March, 1955. A wonderful film.

RISK ASSESSMENT

134. VALENTIN BONDARENKO: From p100-102 of Dominic Phelan [editor] (2013) *Cold War Space Sleuths: The Untold Secrets of the Soviet Space Program*, Springer Praxis, Chichester.

135. A CRATER ON THE MOON IS NOW NAMED AFTER HIM: From *The Red Stuff*, as above.

136. VLADIMIR KOMAROV: Amazing reading by one

of his wives in *The Red Stuff*, as above.

137. GEORGY DOBROVOLSKY, VLADISLAV VOLKOV, AND VIKTOR PATSAYEV: NASA, *Descent into the Void*, as above.

STAYING HEALTHY – A NOTE FROM YOUR DOCTOR

138. DR OLEG KOTOV: by email, 4 June 2017. Translated by Dr Iya Whiteley, Director of the Centre for Space Medicine, Mullard Space Science Laboratory, UCL.

CAN I BUY A TICKET TO SPACE?

139. 'MAY BE OUT OF THIS WORLD': Various sources, e.g. Jeff Gates (2016) *I Was a Card-Carrying Member of the "First Moon Flights" Club*, 20 October. Available at www.smithsonianmag.com/smithsonian-institution/i-was-card-carrying-member-first-moon-flights-club-180960817.

ASTRONAUT WANTED: NO EXPERIENCE NECESSARY

140. 'ASTRONAUT WANTED. NO EXPERIENCE NECESSARY.': Various sources but one is BBC News (2008) *1991: Sharman becomes first Briton in space. On This Day*. Available at news.bbc.co.uk/onthisday/hi/dates/stories/may/18/newsid_2380000/2380649.stm.

141. 'LIKE LISTENING TO AN ORCHESTRA OF COLOURS': From Tomoko Otake (2013) *Toyohiro Akiyama: Cautionary tales from one not afraid to risk all*, The Japan Times, 3 August. Available at www.japantimes.co.jp/life/2013/08/03/people/cautionary-tales-from-one-not-afraid-to-risk-all.

BUYING YOUR OWN TICKET

142. GALL BLADDER STORY: Richard Garriott told me this in the pub.

143. SEVEN PRIVATE, SELF-FUNDED ASTRONAUTS: With grateful thanks to Wikipedia user Rillian, who I believe first compiled this table of information from other sources in June 2014, and other contributors since.

INCENTIVES – THE X PRIZE

144. 'BUILD AND LAUNCH A SPACECRAFT...': From the Ansari X Prize website. Available at ansari.xprize.org.

145. 'IT WAS LIKE BEING HIT WITH A SLEDGE HAMMER...': BBC (2006) *Space Tourists*, Horizon.

VIRGIN GALACTIC – WHITE KNIGHT/ SPACESHIPTWO

146. 'BY TICKING THIS BOX YOU ARE CONFIRMING...': Virgin Galactic (2017) *Future Astronaut Application Form*. Available at www.virgingalactic.com/human-spaceflight/fly-with-us/application.

BALLOON – THE ORBITAL PERSPECTIVE

147. 'COMFORTABLE, STYLISHLY-APPOINTED SPACECRAFT': World View Enterprises, Inc. (2017) *The Experience*. Available at www.worldview.space/voyage.

148. AL WORDEN: Interviewed over Skype, 12 May 2017. Transcribed by Louise Crane.

THIRD STAGE: THE OTHER SIDE OF SKY

149. THE OTHER SIDE OF SKY: The title of a song by Rachel Robinson, from her eponymous album in 2003. A hugely important song in my life. Rachel knows why.

150. 'THERE WAS A TIME WHEN MEN STARED LONGINGLY...': From p87 of Leonard de Vries, *The Second Book Of Experiments*, as above.

ADVENTURES IN LOW EARTH ORBIT

THE INTERNATIONAL SPACE STATION – YOUR HOME AWAY FROM HOME

151. EXPLORING THE INTERNATIONAL SPACE STATION: You can now explore the real thing yourself, starting from the Cupola, via Google Street View. Available at www.google.com/streetview/#international-space-station/cupola-observational-module.

HOW TO JUMPSTART A DEAD RUSSIAN SPACE STATION

152. 'I LOOK AROUND THE STATION...': From the 1 September 1982 diary entry of Valentin Lebadev on Salyut 7 Expedition 1, via p93-4 of David S. F. Portree (1995) *Mir Hardware Heritage*, NASA Reference Publication 1357, March. Available at ston.jsc.nasa.gov/collections/TRS/_techrep/RP1357.pdf.

153. 'ONE OF THE MOST IMPRESSIVE FEATS...': From p99 of David S. F. Portree, *Mir Hardware Heritage*, as above.

154. SPIT WOULD FREEZE TO THE WALLS AND ICICLES HUNG FROM THE PIPES: From p3 of Nickolai Belakovski (2014) *The little-known Soviet mission to rescue a dead space station*, Ars Technica, 16 September. Available at arstechnica.com/science/2014/09/the-little-known-soviet-mission-to-rescue-a-dead-space-station. A superbly written article.

155. 'IT FELT LIKE BEING IN AN OLD, ABANDONED HOME...': Savinykh in his flight journal. From p3 of Nickolai Belakovski, *The little-known Soviet mission to rescue a dead space station*, as above.

156. 'THAT DAY WAS THE FIRST HAPPY SPARK OF HOPE...': Savinykh in his flight journal. From p3 of Nickolai Belakovski, *The little-known Soviet mission to rescue a dead space station*, as above.

CONFLICT

157. 'WE GOT ALONG TOGETHER JUST FINE...': At a post-mission debriefing, as reported by Michael Hiltzik (2015) *The day when three NASA astronauts staged a strike in space*, Los Angeles Times, 28 December. Available at www.latimes.com/business/hiltzik/la-fi-mh-that-day-three-nasa-astronauts-20151228-column.html.

158. SKYLAB 4 STORY: From Mary M. Connors, Albert A. Harrison and Faren R. Akins (1985) Chapter VIII. ORGANIZATION AND MANAGEMENT – External Relations, in *LIVING ALOFT: Human requirements for extended spaceflight*, NASA SP-483, 1 January. Available at ntrs.nasa.gov/search.

jsp?R=19850024459.
159. SPACE MUTINY: Emily Carney summarizes this story from two sources: a 2008 book by David Hitt and others (*Homesteading Space: The Skylab Story*, University of Nebraska Press) and an October 1974 National Geographic article by Thomas Y. Canby (*Skylab, Outpost on the Frontier of Space*, vol. 146. issue 4, 441-469). See Emily Carney (2016), *Space Myths Busted: No, There Wasn't A "Mutiny" On Skylab*, This Space Available, 3 January. Available at this-space-available.blogspot.co.uk/2016/01/space-myths-busted-no-there-wasnt.html.
160. POGUE'S SPACE SICKNESS: From p183 of Bryan Burrough (1998) *Dragonfly: NASA and the Crisis Aboard Mir*, HarperCollins, New York.
161. 'PUT ON YOUR BLINDERS AND HEAD NORTH' From p184 of Bryan Burrough, *Dragonfly: NASA and the Crisis Aboard Mir*, as above.

SHUTTLE–MIR AND THE THIRD MAN

162. 'WHO IS THE THIRD WHO WALKS ALWAYS BESIDE YOU?…': T.S. Eliot (1922), *The Waste Land*. Faber and Faber Ltd.
163. JERRY LINENGER STORY: Reported by Abbie Bernstein (2011), *The Angel Effect – An Interview with Dr. Jerry Linenger*, Buzzy Mag, 26 April. Available at buzzymag.com/the-angel-effect-inteview-with-dr-jerry-linenger.
164. 'IT WAS MY DAD, IT WAS HIM AND I HAVE NO DOUBT…': Interviewed in National Geographic Explorer's *The Angel Effect*, first broadcast 26 April 2011. See also Abbie Bernstein, *The Angel Effect – An Interview with Dr. Jerry Linenger*, as above.
165. MICHAEL FOALE'S FAVOURITE FILMS: These were two of the three films he mentioned in his talk *20th anniversary of the Mir space station collision*, hosted by Pint of Science at the Greenwood Theatre in London, 25 June 2017.
166. 'IT WAS FULL OF MENACE — LIKE A SHARK…': From [32:16] of BBC (1998) *Mir Mortals*, Horizon, first broadcast 23 April 1998. Available at www.bbc.co.uk/iplayer/episode/p02863gw/horizon-19971998-mir-mortals.
167. 'MICHAEL, TO THE SHIP!': Various sources, e.g. [31:38] of BBC, *Mir Mortals*, as above. ('Michael! Va korabl!').
168. 'LIKE A DOG': Michael Foale quoting Sasha, *20th anniversary of the Mir space station collision*, as above.
169. APOLLO 13 FILM STORY: Michael Foale, *20th anniversary of the Mir space station collision*, as above.

END OF LIFE –
SPACE DEBRIS AND THE RETURN TO EARTH

170. 'DEVILLS AND WICKED SPIRITS…': From Part IV of Francis Godwin, *The Man on the Moone*, as above (p88-89 in William Poole edition).
171. 3600 STILL REMAIN UP THERE: European Space Agency reported by Associated Press (2013), e.g. David Rising (2013) *Satellite hits Atlantic—but what about next one?* Phys.org, 11 November. Available at https://phys.org/news/2013-11-satellite-atlanticbut.html.
172. SUITSAT: You can immerse yourself in the 'secret world of space junk' in Nick Ryan and Cath Le Couter's *Project Adrift*, an online project that re-imagines the life of some of the litter now flying above our heads. Available at www.projectadrift.co.uk.
173. 'A PIECE OF THE ABANDONED SOVIET SPACE STATION SALYUT-7…': Various, e.g. p10 of The Canberra Times (1991), *A piece of Salyut-7 crashes on to patio*, 9 February. Available at trove.nla.gov.au/newspaper/article/129095216.

THE SPACE CRAFT CEMETERY.
THE OCEAN POLE OF INACCESSIBILITY

174. 'AH, SIR, LIVE IN THE BOSOM OF THE WATERS!…': From p360 of Jules Verne (1865), *20,000 Leagues Under the Sea*, in: *Around the World in Eighty Days; From The Earth to the Moon Direct; 20,000 Leagues Under the Sea*, Octopus Books Limited, London (this edition 1978).

DESTINATION: MOON

175. 'AFTER THE TIME I WAS ONCE QUITE FREE…': From Part V of Francis Godwin, *The Man in the Moone*, as above (p97 in William Poole edition).

TOURIST INFORMATION

176. 'STARLESS AND BIBLE BLACK': From the opening of Dylan Thomas's *Under Milk Wood*.

GETTING THERE – SATURN 5 AND N1

177. '…IF YOU DRIVE A CAR WITH AN AUTOMATIC TRANSMISSION…': Al Worden interviewed at Alton Towers, 3 October 2016. Transcribed by Tamsin Edwards.
178. CREATING ONE OF THE BIGGEST NON-NUCLEAR EXPLOSIONS EVER: For example see p22 of Michel van Pelt (2017) *Dream Missions: Space Colonies, Nuclear Spacecraft and Other Possibilities*, Springer Praxis, Chichester.

FINDING YOUR WAY –
KNIT YOUR OWN SATNAV

179. 'STAR PATTERNS ARE VERY IMPORTANT…': Al Worden interviewed at Alton Towers, 3 October 2016. Transcribed by Tamsin Edwards.
180. THE MEMORY OF A MUSICAL GREETING CARD: Estimates and comparisons abound. This one uses the brilliantly thorough book by Frank O'Brien (*The Apollo Guidance Computer Architecture and Operation*, Springer), which says on p31 that the core rope read only memory (ROM) was 36,864 15-bit 'words', and a byte is 8 bits, so that's 36,864 words x 15 (bits per word) / 8 (bits per byte) bytes, or 69.1 kilobytes of ROM (along with 4.1 kilobytes of magnetic read/write memory). The idea for comparing with a musical greeting card came from a talk I saw by Dr James Bellini. They only need read only memory, and the amount is about right: in Examples 5.13 and 5.14 of Roman Kuc's textbook *The Digital Information Age: An Introduction to Electrical Engineering*, the musical greeting cards need 30-40 kilobits per second of recording, so the Apollo Guidance Computer's core rope 553 kilobits would give you a decent 13 to 18 seconds.

WHO, WHAT, WHEN AND WHY OF APOLLO MISSIONS

181. APOLLO MISSION NOTES: Several facts courtesy of the excellent Timothy B. Benford and Brian Wilkes (1985) *The Space Program Quiz & Fact Book*, Harper & Row, New York. Thanks for the recommendation Chris.

MAN IN THE MOONE, FRAU IM MOND

182. 'I AM THE VINE…': See p25 of Buzz Aldrin with Ken Abraham (2009) *Magnificent Desolation: The Long Journey Home from the Moon*, Harmony Books, New York.

183. MOON MAIDEN: So argues Françoise Launay from the Observatoire de Paris, in his 2003 letter *The Moon Maiden of Cassini's map*, Astronomy & Geophysics, vol. 44, issue 1, 1.7-1.7, February. Available at adsabs.harvard.edu/full/2003A%26G....44a...7.

TREAD LIGHTLY

184. RECOMMENDATIONS TO SPACE-FARING ENTITIES: NASA (2011), *Recommendations to Space Faring Entities: How to Protect and Preserve the Historic and Scientific Value of U.S. Government Lunar Artefacts*, 20 July. Available at www.nasa.gov/pdf/617743main_NASA-USG_LUNAR_HISTORIC_SITES_RevA-508.pdf.

185. LONG-TERM DEGRADATION: T.W. Murphy Jr. (2010) *Long-term degradation of optical devices on the Moon*, Icarus 208, 31-35.

186. 'PUSH BIGGEST POSSIBLE ROCK…': From p52 of NASA, *Recommendations to Space Faring Entities*, as above.

187. LIST OF ITEMS ON THE MOON: From p82-85, Apollo 15 table, in Appendix E of NASA, *Recommendations to Space Faring Entities*, as above.

LUNAR XPRIZE

188. 'SUCCESSFULLY PLACE A SPACECRAFT ON THE MOON'S SURFACE…': From the website Lunar.xprize.org.

189. 'WE CHOOSE TO GO TO THE MOON…': From Kenneth Chang (2013), *Florida Company Gets Approval to Put Robotic Lander on Moon*, New York Times, 3 August. Available at www.nytimes.com/2016/08/04/science/moon-express-faa.html.

190. AMERICAN MOON EXPRESS TEAM: From Kenneth Chang, *Florida Company Gets Approval to Put Robotic Lander on Moon*, as above.

191. WE HAVE LEFT TONNES OF STUFF ON THE MOON: A widely cited number is 187,400 kilograms, e.g. p8 of Jamie Carter (2016) *The Moon is a tech museum*, Tech Radar, 5 March. Available at www.techradar.com/news/world-of-tech/the-moon-is-a-tech-museum-1316285.

192. JAVELIN CRATER: A photo of the crater, javelin and ball is at www.hq.nasa.gov/alsj/a14/a14det9337.jpg, which is a detail of AS14-66-9337 described in the Apollo 14 Image Library at www.hq.nasa.gov/alsj/a14/images14.html#9337.

ART ON THE MOON

193. 'MY HOMO SAPIENS, MY CYBERNETIC MAN…': From p30 of William Wertenbaker (1972) *Only Artist on the Moon*, The Talk of The Town, The New Yorker, 20 May.

194. FALLEN ASTRONAUT STORY: For more see Rick Mulheirn and Danny Van Hoecke (2015) *Honour to the Fallen Astronauts*, Spaceflight, The British Interplanetary Society, vol. 57, no. 10, 382-388, October.

195. 'LARGEST EXHIBITION SPACE…': Corey S. Powell and Laurie Gwen (2013), *The Sculpture on the Moon*, Slate, 16 December. Available at www.slate.com/articles/health_and_science/science/2013/12/sculpture_on_the_moon_paul_van_hoeydonck_s_fallen_astronaut.html.

ANDY WARHOL'S TINY PENIS

196. WHY WOULD AN ENGINEER RISK EVERYTHING: PBS (2010) *Moon Museum*, Episode 1, Season 8 of History Detectives, PBS, broadcast 21 June. Video available at www.youtube.com/watch?v=ppJGiw6mCxk. Transcript available at www.pbs.org/opb/historydetectives/investigation/moon-museum.

197. THE MYSTERY WAS INVESTIGATED: PBS, *Moon Museum*, as above.

BACK TO THE MOON

198. 'MAGNIFICENT DESOLATION': Buzz Aldrin's book, as above, named after his first impression of the moon.

199. 'THE MOON'S RICHES…': Around [57:28] of Fritz Lang, *Frau im Mond*, as above. Available at www.youtube.com/watch?v=aHcazI9PgNg.

200. 'SOLVE THE WORLD'S BIGGEST PROBLEMS': From [1:32] of Naveen Jain interview within article Lori Ioannou (2017), *Billionaire closer to mining the moon for trillions of dollars in riches*, CNBC, 31 January. Available at www.cnbc.com/2017/01/31/billionaire-closer-to-mining-moon-for-trillions-of-dollars-in-riches.html.

STONED MOON

201. YOUR COUNTRY'S MOON ROCKS: A list of lunar displays is at curator.jsc.nasa.gov/lunar/displays/index.cfm#history.

202. THAD SPENT AN OUT-OF-THIS-WORLD NIGHT: Ben Mezrich (2011) *Sex on the Moon: The Amazing Story Behind the Most Audacious Heist in History*, DoubleDay, New York.

MOON VEXILLOLOGY

203. 'AS FOR THE YANKEES…': From p213 of Jules Verne, *From the Earth to the Moon*, as above.

204. KINZLER STORY: From Sandra L. Johnson (2008) *Red,White, & Blue: U.S. Flag at Home on the Moon*, Houston History, vol. 6, no. 1, Fall. Available at houstonhistorymagazine.org/wp-content/uploads/2014/03/red-white-and-blue-US-flag.pdf.

205. 'HERE MEN FROM THE PLANET EARTH…': See for example NASA (2017) *July 20, 1969: One Giant Leap For Mankind*, 20 July. Available at www.nasa.gov/mission_pages/apollo/apollo11.html.

WHOSE FLAG IS IT ANYWAY?

206. 'THE STARS AND STRIPES TO BE DEPLOYED ON THE MOON …': NASA Press Release 69-83E, 3 July 1969 via Anne Platoff (1993) *Where No Flag Has Gone Before: Political and Technical Aspects of Placing a Flag on the Moon*, NASA Contractor Report 188251, August. Available at ntrs.nasa.gov/

archive/nasa/casi.ntrs.nasa.gov/19940008327.pdf.

207. FLAG MYSTERY: Anne Platoff has done much of the heavy lifting in *Where No Flag Has Gone Before*, as above. Jeremy Markovich looked into this more recently in his 2016 article *Apollo 11, an American Flag, a Small Town, and a Mystery*, Our State: Celebrating North Carolina, June. Available at www.ourstate.com/rhodiss-american-flag-mystery. See also Annin's version of the story at Marian Calabro (2013) *Annin Flagmakers: An Illustrated History*. Available at www.annin.com/downloads/Annin_History_Book.pdf.

208. 'ANNIN HAS PRODUCED ALL THE FLAGS…': From p4 of Rotary International (1971), *The Rotarian*, vol. 118, no. 6, June.

209. ANNIN HAVE CONFIRMED TO ME: From Mary E. Repke (2017), Annin Senior Vice President of Sales and Marketing, by email, 5 January.

LET'S GO TO MARS

210. 'YET ACROSS THE GULF OF SPACE…': From p1 of H.G. Wells (1898) *The War of the Worlds*.

MARTIANS WANTED

211. 'MEN WANTED FOR HAZARDOUS JOURNEY…': See for example Colin Schultz (2013) *Shackleton Probably Never Took Out an Ad Seeking Men for a Hazardous Journey*, Smithsonian Magazine, September. Available at www.smithsonianmag.com/smart-news/shackleton-probably-never-took-out-an-ad-seeking-men-for-a-hazardous-journey-5552379.

212. OPPOSITION-CLASS AND CONJUNCTION-CLASS: Estimates and diagrams from Bryan Mattfeld and others (2014) *Trades Between Opposition and Conjunction Class Trajectories for Early Human Missions to Mars*, Space, San Diego, United States, 4-7 August. Available at ntrs.nasa.gov/search.jsp?R=20150001240.

A STRANGER IN A STRANGE LAND

213. WAR OF THE WORLDS STORY: From Martin Chilton (2016) *The War of the Worlds panic was a myth*, The Telegraph, 6 May. Available at www.telegraph.co.uk/radio/what-to-listen-to/the-war-of-the-worlds-panic-was-a-myth. See also Jefferson Pooley and Michael Socolow (2013) *The Myth of the War of the Worlds Panic*, Slate, 28 October. Available at www.slate.com/articles/arts/history/2013/10/orson_welles_war_of_the_worlds_panic_myth_the_infamous_radio_broadcast_did.html.

214. 'FAKE RADIO WAR STIRS TERROR THROUGH U.S.': In the *New York Daily News* (1938), 31 October, according to Jefferson Pooley and Michael Socolow, *The Myth of the War of the Worlds Panic*, as above.

ADJUSTING THE FOCUS

215. 'ALL PHILOSOPHY, SAID I, IS FOUNDED ON TWO THINGS…': From Bernard le Bouvier de Fontenelle (1686), *First Evening* from *Conversations on the Plurality of Worlds*. Translated by Miss Elizabeth Gunning (1803).

216. 'I HAVE OFTEN NOTICED OCCASIONAL CHANGES…': From the final page of William Herschel (1784), *On the remarkable Appearances at the Polar Regions of the Planet Mars, the Inclination of its Axis, the Position of its Poles, and its spheroidical Figure; with a few Hints relating to its real Diameter and Atmosphere*, Philosophical Transactions of the Royal Society of London, vol. 74, 233-273.

LOST IN TRANSLATION

217. ALCOHOL, NARCOTICS AND COFFEE: From p84 of Edward S. Morse (1906), *Mars and its Mystery*, Little, Brown and company, Boston. Available at archive.org/details/marsitsmystery00mors.

218. 'WHERE WE HAVE STRONG EMOTIONS…': Carl Sagan (1977), *Royal Institution Christmas Lectures*, Royal Institution of Great Britain.

GETTING YOUR ASS TO MARS

219. 'IT'LL BE REALLY FUN TO GO. YOU'LL HAVE A GREAT TIME': From [42:29] of Elon Musk (2016), *Making Humans a Multiplanetary Species*, 67th International Astronautical Congress, 27 September. Available at www.spacex.com/mars.

220. 'WHERE THERE'S A WILL THERE'S A WAY…': From [15:35] in Robert Zubrin (2017) *The Tools for the Task*, On the Launchpad: Return to Deep Space, Atlantic Live. 16 May 2017. Available at www.youtube.com/watch?list=PLwj46yNDLyTU0_Mk58F72t0urwIjgrsw-&v=iZJKAqSZnA4.

BEYOND…

10 EXCITING PLACES TO VISIT IN THE UNIVERSE

221. 'ALL THESE WORLDS ARE YOURS – EXCEPT EUROPA…': Arthur C. Clarke (1982), *2010: Odyssey Two*.

222. DR LOUISA PRESTON: By email, April-June 2017.

ENVOYS: EXTENDING OURSELVES

223. 'THE ONLY TRUE VOYAGE OF DISCOVERY…' From Marcel Proust (1923) *La Prisonnière* from *À la recherche du temps perdu*, translated by CK Scott Moncrieff. Misquotes of this abound.

MARINER 4

224. MARINER 4 STORY: From Blaine Baggett's 2013 documentary *The Changing Face of Mars*, of which the Mariner 4 part is shown in the second half of NASA (2015), *1965: Discovery at Mars*, von Kármán lecture series, first broadcast 16 July. Published online by NASA Jet Propulsion Laboratory, 20 June 2015. Available at www.youtube.com/watch?v=q5dzDWjN7Z4.

225. $3.8 MILLION A PICTURE: Dividing the mission cost of $83.2 million by 22 pictures. Mission cost from NASA (2017), *Mariner 4*, 21 March. Available at nssdc.gsfc.nasa.gov/nmc/spacecraftDisplay.do?id=1964-077A.

226. NEARLY THREE WEEKS TO TRANSMIT: From NASA, *Mariner 4*, as above.

227. A HOST OF OTHER SCIENTIFIC EQUIPMENT: From NASA, *Mariner 4*, as above.

228. BETWEEN 9846 AND 12,000 KM: Andrew LePage (2015), *Mariner 4 to Mars*, Drew Ex Machina, 14 July. Available at www.drewexmachina. com/2015/07/14/mariner-4-to-mars

PAINTING BY NUMBERS
229. PAINTING BY NUMBERS STORY: See the excellent article by Dan Goods, *First TV image of Mars: Interplanetary color by numbers*. Available at www.directedplay.com/first-tv-image-of-mars. See also from around 59:00 in NASA, 1965: *Discovery at Mars*, as above.

MESSAGES IN BOTTLES
230. COMMEMORATIVE PENNANT: Anatoly Zak (2014) *Zond-2: An early attempt to touch Mars?* Russian Space Web, 20 November. Available at www.russianspaceweb.com/3mv_Zond-2.html.
231. 'WHOEVER IS INHABITING EARTH IN THAT DISTANT EPOCH...': From the LAGEOS press release, 4 May 1976. See NASA (2016) *Message to the Future*, 2 May. Available at lageos.cddis.eosdis.nasa.gov/Design/Message_to_the_Future.html. See also Appendix A of Carl Sagan (1978) *Murmurs of Earth: The Voyager Interstellar Record*, Random House, USA.
232. THERE WAS ALSO CRITICISM: From p59 of Carl Sagan, *Murmurs of Earth*, as above.

THE FURTHEST WE'VE EVER SAILED
233. 'I MUST GO DOWN TO THE SEAS AGAIN...': From John Masefield, Sea Fever. Reproduced with permission of The Society of Authors as the Literary Representative of the Estate of John Masefield.

HELLO FROM THE CHILDREN OF PLANET EARTH
234. 'HELLO FROM THE CHILDREN OF PLANET EARTH': is one of the greetings on the Golden Record.
235. 'I GET A GASP OF SURPRISE AT TWO NUMBERS THAT I GIVE OUT ...': Jon Lomberg, via Chris Riley.
236. JOHN CASANI APPROACHED CARL SAGAN: Carl Sagan, *Murmurs of Earth*, as above.
237. A HUNDRED AND TWENTY-TWO IMAGES: From p12 of Carl Sagan, *Murmurs of Earth*, as above.
238. 'PAZ E FELICIDADE A TODOS': Peace and happiness to all, spoken by Janet Sternberg on the Golden Record.
239. '"WILL THEY LIKE MY MUSIC ON VENUS?..."': Paraphrased in Dallas Campbell and Chris Riley (2012) *Voyager: the space explorers that are still boldly going to the stars*, The Observer, 21 October. Available at www.theguardian.com/science/2012/oct/21/voyager-mission-leave-solar-system. Original interview from BBC Radio 4 (1983), *Music From A Small Planet*, produced for BBC Scotland by Martin Goldman and R. Carey Taylor, broadcast 29 July.
240. 'IMAGINED SELF-IMPORTANCE': From Carl Sagan's famous 'Pale Blue Dot' speech at Cornell University, 13 October 1994.
241. 'MURMURS OF EARTH': The name of Carl Sagan's book, as above.
242. 'I CAN SENSE STARS...': NASA Voyager 2 Twitter account, by Dr Paul Filmer. Shown in an online

extra for BBC4 (2012) *Voyager: To the Final Frontier*, 24 October. Clip available at www.bbc.co.uk/programmes/p00zwbl7.
243. JAN WOERNER: Interviewed over Skype, 8 May 2017. Transcribed by Louise Crane.

DESPERATE MEASURES

HOW TO GET ABDUCTED BY ALIENS
244. FIVE POINT PLAN ON HOW TO LEAVE THE PLANET: See Douglas Adams, *Ultimate Hitchhiker's Guide* edition for this lovely bit.

HERE I AM
245. 'IMAGINED SELF-IMPORTANCE': Carl Sagan, as above.

LANGUAGE
246. 'THE DIFFICULTY OF THAT LANGUAGE IS NOT TO BEE CONCEIVED...': From Part VIII of Francis Godwin, *The Man in the Moone*, as above (p108 of William Poole edition).

CODA
247. 'A YOUNG AND IMPRESSIONABLE MOTH ONCE SET HIS HEART ON A CERTAIN STAR...': *From The Moth and The Star* from Fables For Our Time & Famous Poems. Illustrated by James Thurber. Copyright ©1940 by Rosemary A. Thurber. Reprinted by arrangement with Rosemary A. Thurber and The Barbara Hogenson Agency. All rights reserved. Can be found in *The Thurber Carnival*, one of my favourite books I studied at school.
248. SHALL WE GO? YES, LET'S GO: One for the Beckett fans. How appropriate that the last words of *Waiting for Godot* were also Yuri Gagarin's first words as he left the earth. <They do not move>.

RETURN TO EARTH
249. 'NATURE, THAT FRAM'D US OF FOUR ELEMENTS...': The closing words of *The Restless Sphere: The Story of International Geophysical Year*, the wonderful documentary by the BBC and Royal Society presented by HRH Prince Phillip The Duke of Edinburgh, first broadcast 30 June 1957. Details at www.bbc.co.uk/programmes/p00yt7l6. Uploaded online by Stephen Smith, 13 May 2016. Available at www.youtube.com/watch?v=MD5ale35jc4.

Above: **Man Above Planet, Alexei Leonov (1968)**

PICTURE CREDITS

1. Everett Collection/Alamy
2. NASA/Johnson Space Center
3. Old Visuals/Alamy
6. Douglas Pulsipher/Alamy
8. NASA/Alamy
9. Author's own collection
11. AGE Fotostock/Alamy
15. Pablo Carlos Budassi licensed under CC BY-SA 3.0
16. Getty Images
17. Science & Society Picture Library/Getty Images
19. Agnes Meyer-Brandis, VG-Bild Kunst 2017
23. Author's own collection
24. ESA Images
25. Science Photo Library/Alamy
29. Fisher Space Pen Company
30. Soviets in Space/Peter L Smolders, Lutterworth Press
31. RGB Ventures/Superstock/Alamy
32. Tass/Barcroft Media
33. Jonas Bendiksen/Magnum photos
34. (1) Itar-Tass Photo Agency/Alamy, (2) Science History Images/Alamy, (3) Getty Images/Keystone-France, (4) Everett Collection/Alamy, (5) Bettemann/Getty Images
36. Collection Christophel/Alamy
37. Collection Christophel/Alamy
38. (top) Eureka Entertainment, (bottom) NASA/Dimitri Gerondidakis
39. Everett Collection/Alamy
41. Wolfgang Winter/Alamy
42–43. Tom Lehrer/Maelstrom Music
44. NASA Images
45. NASA Images
48. Interfoto/Alamy
49. NASA/MSFC
51. Granger Historical Picture Archive/Alamy
52. (top) Science Photo Library, (bottom) Science & Society Picture Library/Getty Images
54. DBI Studio/Alamy
55. Itar-Tass Photo Agency/Alamy
57. Time & Life Pictures/Getty Images
59. (top) Duncan P Walker/Getty Images, (bottom) Gérard Chatelier
60. 2014 New York Post/Getty Images
62. FLPA/Rex/Shutterstock
63. Interfoto/Alamy
64–67. Agnes Meyer-Brandis, VG-Bild Kunst 2017
68. Science History Images/Alamy
69. Jim Urquhart/Reuters
70–71. NASA Images
72. ESA/Mars 500 Crew
73. Alexis Rosenfeld/Divergence-Images
75, 78, 81. Beth Healey
84. Michael Cockerham
89. NASA Images
90. (background) Sirin Bird/Shutterstock, (left) Wright Air Development Center
92. Chronicle/Alamy
93. Tamsin Edwards
94. World History Archive/Alamy
96. NASA Images
97. NASA/Johnson Space Center
99. RGB Ventures/Superstock/Alamy
101. Cristina De Middel
103. (top left) NASA Images, (top right) Mondadori Collection/Getty Images, (bottom) NASA Images
105. Time & Life Pictures/Getty Images
106. Author's own collection
107. Mary Evans Picture Library
108. Author's own collection
109. (top left) Herge/Moulinsart 2017, (top right) FPG/Getty Images
111. John Frost Newspapers/Alamy
113. (left) Itar-Tass Photo Agency/Alamy (right) Author's own collection
114–5. AP/Rex/Shutterstock
116–7. Author's own collection
118-9. Getty Images, ESA Images, NASA Images
120. The Boeing Company
121. (left) Polaris/Eyevine, (right) Bill Stafford-NASA Johnson Space Center
125. (top) Heritage Image Partnership LTD/Alamy, (bottom left) NASA/Johnson Space Center, (bottom right) Ray Cunningham
126. Granger/Rex/Shutterstock
127. NASA Images
129. Thorsten Schmidt
132. Producti/Rex/Shutterstock
133. (top) 2016 Jeffrey Gates. Original in the collection of the Smithsonian National Air and Space Museum, (bottom left) Broker/Alamy, (bottom right) Pan Am Systems
134. Itar-Tass Photo Agency/Alamy
138. Keystone USA-ZUM/Rex/Shutterstock
139. Blue Origin/Alamy

140. BBC Newsround
141. NASA Images
142. Aviation History Collection/Alamy
143. Alun Reece/Alamy
145, 149, 153. NASA Images
154–55. ESA Images
156. Sputnik/Alamy
157. ESA/Columbus Control Centre
160. Space Agency/Rex/Shutterstock
161. Rob Elliott
162. Cindy Hopkins/Alamy
162–3. Agnes Meyer-Brandis, VG-Bild Kunst 2017
164. Collection Christophel/Alamy
166. NASA Images
167. Nikolai Ignatiev/Alamy
168, 169. Raytheon via David Meerman Scott
171. Science Photo Library
172. Julie Libersat, *First Manmade Object Lands on the Moon*, (published September 12, 2011 at timesillustrated.com). Courtesy of the artist
173. Images GRO/Rex/Shutterstock
174. NASA Images
176. Courtesy of Paul Van Hoeydonck
177. NASA Images
178. Moon Museum, The Museum Of Modern Art, New York/Scala, Florence
179, 182–3, (top) 185. NASA Images
185. (bottom) Heritage Image Partnership LTD/ Alamy
186–8. NASA Images
189. The National Archives
190. SpaceX
191, 192. NASA Images
193. Bryan Mattfeld, Chel Stromgren, Hilary Shyface, David Komar, William Cirillo and Kandyce Goodliff/American Institute of Aeronautics and Astronautics
194. SpaceX
196. Science Photo Library
197. (top left) History Archive/Rex/Shutterstock, (top right) SIPA Press/Rex/Shutterstock, (bottom) Britannica/UIG/Rex/Shutterstock
198. Science Photo Library
200. SpaceX
201. Science History Images/Alamy
206. Spencer Lowell
207, 208–9. NASA/Dan Goods/JPL
210. Granger/Rex/Shutterstock
211. NASA Images

212–3. Granger Historical Picture Archive/Alamy
214, 215. NG Images/Alamy
216. International Music Score Library Project licensed by CC BY-SA 4.0
217. NASA Images
222. Otis T Carr/U.S 2912244
223. Erin Paul Donovan/Alamy
225. Lance Purple
227. Elysium
228. AFP/Getty Images
231. Paul Fearn/Alamy
233. NASA Images
235. Collection Christophel
247. Sputnik/Alamy
Endpapers. Ronald M Jones/ integratedspaceanalytics.com

INDEX

ACKNOWLEDGEMENTS

This book is a collection of other people's research, experiences and anecdotes, which I've had the pleasure of stitching together over the last year. There are many to whom I owe a deep debt of gratitude.

Thank you to everyone at BBC Science, led by Andrew Cohen, for the extraordinary opportunities that working with you has given me. To my friends Helen Czerski, Jim Al-Khalili, Jo Durrant and everyone in the Cheltenham Science Festival green room. To the great Emma Pound for letting me copy your homework. To Adam Rutherford (as mentioned in Viz) and Georgia Murray for the moral support and the many splendid evenings.

Thank you Gill Norman for letting me wild camp in the British Interplanetary Society Library. To Richard Garriott for comparing scars in the pub. Art Dula and the Heinlein Prize Trust for our spacesuit adventures. Libby Jackson and everyone at the UK Space Agency, Sheona Urquhart for the whistle stop tour of the universe. Roger Highfield, Doug Millard and the Science Museum staff for such valued friendship, and to Helen Sharman who I always enjoy bumping into there. To George Abbey for regaling me with early wild west NASA stories over breakfast.

Thank you to my cherished agents, Tracey McLeod, Theia Nankivell, Chloe Gott and everyone at KBJ mission control. Thanks to the ground crew: Louise Crane, George Perry (the fact checker's fact checker), Mike Jones for saving me from myself, my superb editor Nicki Crossley for constantly reminding me what the book was about when I'd forget, and all at Simon & Schuster who have put up with me and my schedule with patience and good humour. To Ian Whent for photograph chasing, Bee Willey for the lovely illustrations and Ash Western for making this book look so beautiful and going many, many extra miles.

Thank you to the legions of contributors: Thorsten Schmidt, Jeremy Markovich for our moon flag chat, Louisa Preston, The Arts Catalyst, Paul and Marleen Van Hoeydonck, Agnes Meyer-Brandis, Gérard Chatelier, David Meerman Scott, Rob Elliott, Cristina de Middel and Rob La Frenais. Everyone at ESA, particularly Jan Woerner, Margherita Buoso, Maxim Mommerency and Romain Charles who shared with me the fascinating story of the Chilean miners. Thanks Al Worden – a true American hero, and to Vix Southgate for our day at Alton Towers. Thank you Jill Stuart and Oleg Kotov for your advice, Iya Whiteley for the translating and the origami lessons, and Tom Lehrer for your charming e-mail. Also Lori Styler and Rosemary Thurber for your eleventh hour kindness and permission to reprint The Moth and The Star. Thanks to Beth Healey for letting me tag along on your adventures. To Tim and Rebecca Peake, Michael Foale and all the astronauts who have kindly signed my Ladybird Exploring Space book.

A special thanks to filmmaker Christopher Riley, whose depth of knowledge is remarkable – thank you for my favourite space fact: Michael Collins, en route to the Apollo 11 launch, can be seen holding a mysterious large brown paper bag. It contained a small trout – a gift for the launchpad 'czar', Guenter Wendt.

Apologies to my family for putting up with me spending a year off-planet with the many communication blackouts. To William Sutton for a lifetime of friendship and inspiration.

Finally to Tamsin Edwards, whose creativity, thoroughness and work ethic is without equal – thank you for the structural support, IT support and life support. And without whom, this book wouldn't exist.

DALLAS CAMPBELL
July 2017

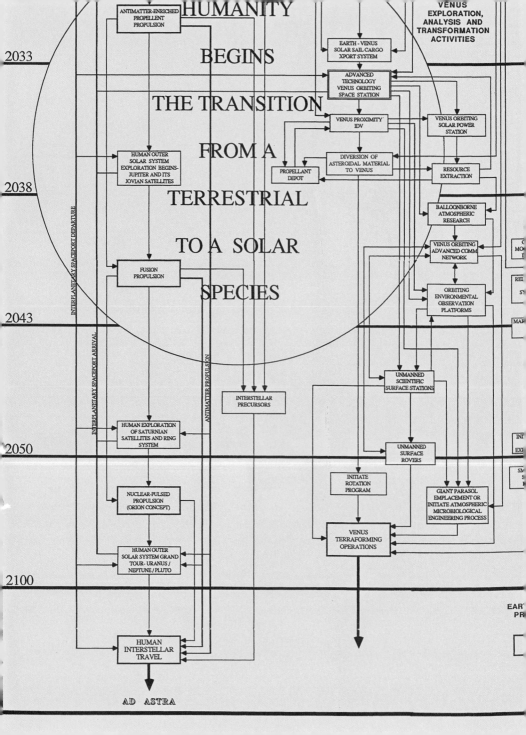

HUMANITY

BEGINS

THE TRANSITION

FROM A

TERRESTRIAL

TO A SOLAR

SPECIES

VENUS
EXPLORATION,
ANALYSIS AND
TRANSFORMATION
ACTIVITIES

2033

2038

2043

2050

2100

ANTIMATTER-ENRICHED
PROPELLENT
PROPULSION

EARTH - VENUS
SOLAR SAIL CARGO
XPORT SYSTEM

ADVANCED
TECHNOLOGY
VENUS ORBITING
SPACE STATION

VENUS PROXIMITY
IDV

VENUS ORBITING
SOLAR POWER
STATION

HUMAN OUTER
SOLAR SYSTEM
EXPLORATION BEGINS-
JUPITER AND ITS
JOVIAN SATELLITES

DIVERSION OF
ASTEROIDAL MATERIAL
TO VENUS

PROPELLANT
DEPOT

RESOURCE
EXTRACTION

BALLOONBORNE
ATMOSPHERIC
RESEARCH

VENUS ORBITING
ADVANCED COMM
NETWORK

FUSION
PROPULSION

ORBITING
ENVIRONMENTAL
OBSERVATION
PLATFORMS

INTERPLANETARY SPACEPORT DEPARTURE

INTERPLANETARY SPACEPORT ARRIVAL

ANTIMATTER PROPULSION

INTERSTELLAR
PRECURSORS

UNMANNED
SCIENTIFIC
SURFACE STATIONS

HUMAN EXPLORATION
OF SATURNIAN
SATELLITES AND RING
SYSTEM

UNMANNED
SURFACE
ROVERS

NUCLEAR-PULSED
PROPULSION
(ORION CONCEPT)

INITIATE
ROTATION
PROGRAM

GIANT PARASOL
EMPLACEMENT OR
INITIATE ATMOSPHERIC
MICROBIOLOGICAL
ENGINEERING PROCESS

HUMAN OUTER
SOLAR SYSTEM GRAND
TOUR- URANUS /
NEPTUNE / PLUTO

VENUS
TERRAFORMING
OPERATIONS

HUMAN
INTERSTELLAR
TRAVEL

AD ASTRA